Quantum
Physics

for
dummies®
A Wiley Brand

Quantum Physics

3rd Edition

by Andrew Zimmerman Jones

A Wiley Brand

Quantum Physics For Dummies®, 3rd Edition

Published by: **John Wiley & Sons, Inc.**, 111 River Street, Hoboken, NJ 07030-5774, www.wiley.com

Copyright © 2024 by John Wiley & Sons, Inc., Hoboken, New Jersey

Media and software compilation copyright © 2024 by John Wiley & Sons, Inc. All rights reserved.

Published simultaneously in Canada

For general information on our other products and services, please contact our Customer Care Department within the U.S. at 877-762-2974, outside the U.S. at 317-572-3993, or fax 317-572-4002. For technical support, please visit https://hub.wiley.com/community/support/dummies.

Wiley publishes in a variety of print and electronic formats and by print-on-demand. Some material included with standard print versions of this book may not be included in e-books or in print-on-demand. If this book refers to media such as a CD or DVD that is not included in the version you purchased, you may download this material at http://booksupport.wiley.com. For more information about Wiley products, visit www.wiley.com.

Library of Congress Control Number: 2024933516

ISBN 978-1-394-22550-7 (pbk); ISBN 978-1-394-22551-4 (ePUB); ISBN 978-1-394-22552-1 (ePDF)

SKY10070987_032724

Contents at a Glance

Table of Contents

Introduction

Physics as a general discipline has no limits; it encompasses physical phenomena from the very huge (galaxy-wide) to the very small (atoms and smaller). This book is about the very small side of things — that's the specialty of quantum physics. When you *quantize* something, you can't go smaller; you're dealing with the tiniest discrete units of matter.

Classical physics is terrific at explaining the science behind activities such as heating cups of coffee, accelerating down ramps, or colliding vehicles (as well as a million other things), but it has problems when physical matter gets very small. Quantum physics usually deals with the micro world and examines activities such as what happens when you look at individual electrons zipping around in an atom. And when you get to that tiny level, goings-on can become very strange.

Quantum physics contains principles of uncertainty that affect physicists' ability to precisely identify a particle's physical characteristics. For example, you can't know (with perfect accuracy) a particle's exact position and its momentum at the same time. Quantum physics also explains the way that the energy levels of the electrons bound in an atom work. As physicists probed ever deeper for a way to model reality, quantum physics allowed them to figure out more about this microscopic realm of matter and energy. You encounter all of these topics, and more, in this book.

About This Book

This book contains the need-to-know concepts of quantum physics, including the history of how it was discovered, explanations (and thought experiments) outlining its core issues and debates, and the math needed to dive into some of its central problems.

Quantum physics is one of the most conceptually confusing subjects known to science. It's a subject about which there can be heated, fundamental debates among even the most distinguished experts. The first several chapters focus on exploring the underlying science, how quantum physics ideas were discovered, and the core scientific concepts.

But ultimately, quantum physics is about solving equations, and this book doesn't shy away from that. Starting with Chapter 7, this book assumes a fairly high level of mathematical proficiency. You can't fully appreciate the subject without getting into calculus, and beyond that into subjects such as linear algebra and differential equations.

As you read through the book, you find that I use the following conventions:

>> **Italics** indicate a technical term related to quantum physics, usually the first time it appears. The italicized term is closely followed by its definition or explanation. When that term crops up in later chapters, it isn't necessarily italicized.

>> **Numbered steps** — when working through problems in this book — serve to help organize the process of solving them.

To make the content more accessible, I divided it into five parts:

>> **Part 1: Getting Started with Quantum Physics:** In this part, find out about the basic principles of quantum physics, the physical concepts that led up to them, and the history of their discovery and refinement.

>> **Part 2: The Fundamentals — Quantum Physics Principles and Theories:** This part helps you dig more deeply into the concepts that are at the core of quantum physics, from wave-particle duality and the various types of physical particles, to the possible interpretations of quantum physics.

>> **Part 3: By the Numbers — Basic Quantum Physics Math:** In this part, discover the mathematics that underlies the discoveries of quantum physics, including the notations that physicists have adopted in this field. Then, using that understanding, walk through some of the more direct problems that you can solve with quantum physics.

>> **Part 4: Going 3D with Quantum Physics Calculations:** This part directs you to tackle more complicated problems in quantum physics, including using rectangular and spherical coordinates to approach these problems. You also begin to explore the hydrogen atom and multiple subatomic particles that interact with each other.

>> **Part 5: The Part of Tens:** Enjoy this part as you discover more about some of the key figures and triumphs of quantum physics.

Foolish Assumptions

I don't assume that you have any knowledge of quantum physics when you start to read this book. However, I do make the following assumptions that you

>> Are taking a college-level course in quantum physics, or are interested in how math describes the motion and energy of matter on the atomic and suba-tomic scales.

>> Have some math prowess to understand the content that starts in Chapter 7. In particular, you know calculus and trigonometry, and how to refer to a table of differential equations. You also have some experience with linear algebra in describing a Hilbert space.

>> Have some classical physics background. If you've had a year of college-level physics (or know everything from *Physics For Dummies*), then you should have a solid base of foundational concepts.

Icons Used in This Book

Throughout this book, icons in the margins highlight certain types of valuable information that call out for your attention. Here are the icons you encounter and a brief description of each.

The Tip icon marks tips and shortcuts that you can use to make quantum physics easier.

The Remember icon marks the information that's especially important to know. To siphon off the most important information in each chapter, just skim through these icons.

The Technical Stuff icon marks information of a highly technical nature that you can normally skip over.

The Warning icon tells you to watch out! It marks important information that may save you headaches, mostly related to common misconceptions about quantum physics.

Beyond the Book

In addition to the abundance of information and guidance related to quantum physics that I provide in this book, you get access to even more help and information online at Dummies.com. Check out this book's online Cheat Sheet; just go to www.dummies.com and search for "Quantum Physics For Dummies Cheat Sheet."

Where to Go from Here

This book isn't intended as a linear read, so you can jump around in the content as needed. If you're interested in quantum physics but don't have a strong science background, start at the very beginning in Chapter 1 and work your way through. If you feel you have a solid grasp on classical physics, and want to focus only on the quantum stuff, then you can skip Chapter 2. If you've got a pretty good handle on the conceptual elements of quantum physics and really want to see how to work with the math, then grab your differential equation tables and jump straight to Chapter 7.

And if you're interested in a specific topic that I don't mention in the previous paragraph, you can look for it in the (front of book) *Table of Contents* or the (back of book) *Index.*

1
Getting Started with Quantum Physics

IN THIS PART . . .

Find out about the basic concepts at the heart of quantum physics.

Examine underlying classical physics concepts.

Dive into the major experiments and discoveries of quantum physics.

Chapter **1**

What Is Quantum Physics, Anyway?

Throughout the twentieth century, quantum physics transformed our world. Humanity went from a species that questioned whether atoms existed to one that harnessed the power of the atom. Humans also used the understanding of atoms and subatomic particles to create desktop computers out of microscopic transistors — feats made possible by quantum physics.

Homes across the globe are powered by streams of electrons, a subatomic particle that had been discovered but was barely understood before the rise of quantum physics. The first computers stored information on physical cards, but then transitioned to using magnetism, on both hard drives and floppy diskettes. For a time, humanity stored information on compact disks that machines read by using lasers (also a product of quantum physics). Now, a common storage medium is a solid-state drive (SSD) built of semiconductors (another quantum physics outcome), and most people carry micro supercomputers (called smartphones) in their pockets.

In this chapter, I provide a high-level discussion of the transformation from the classical to quantum understanding of matter and energy. I talk briefly about the world before the discovery of quantum physics and then introduce the key features that physicists discovered when they first began exploring the quantum nature of reality. I discuss why people don't see these quantum effects in their everyday lives, and how improvements in technology allow them to first see and then expand on these understandings to grow their knowledge of physics.

The Classics: Pre-Quantum Physics

At its core, *physics* is the scientific study of the fundamental elements of physical reality: matter and energy. As you scale up the physical structure, and this matter and energy takes the form of chemicals mixed together or, say, a squirrel, the scientific study becomes chemistry and biology. But if you're talking about the baseline study of matter and energy, that means you're talking about physics.

Any scientific field, of course, has a lot of sub-disciplines. If you are studying the squirrel, for example, you aren't just studying biology, but also *zoology* (the study of animals). If you're studying how an acorn grows into a tree, then that would be biology but also *botany* (the study of plants).

If you're studying the path of an acorn hurled from a tree by an angry squirrel, well, that's physics. But it's also the specific field of kinematics. Physics includes many sub-fields, including *thermodynamics* (the study of heat energy), *optics* (the study of light), and *electromagnetism* (the study of electricity and magnetism).

REMEMBER

Throughout this book, I assume that you, the reader, have a general understanding of the basic ideas of classical physics. Chapter 2 focuses on many of the disciplines of classical physics that study matter and energy in different forms and structures. And as these studies became more detailed, they left questions that laid the foundation, at the end of the nineteenth century, for the discovery of an entirely new field of physics — quantum physics.

What Makes Physics Quantum?

Quantum physics refers to a series of discoveries from the first half of the twentieth century and the scientific explanations related to those discoveries. The insights from these explanations revolutionized the understanding of matter and energy at the smallest scale and caused a transformation in the fundamental way that physicists describe and think about the physical reality of these structures. I cover these revolutionary discoveries and experiments in detail in Chapter 3.

But what are the key insights that make quantum physics different from the physics that came before it? A couple of major differences are central to understanding how quantum physics differs from classical physics.

>> **Quantization:** Physical quantities are measured in discrete units, packets, or *quanta* that cannot be broken down any further.

>> **Uncertainty and probability:** Systems have inherent uncertainty built into them.

The bulk of this book explores how these two ideas interact with each other and show up in quantum physical systems, and the implications that arise from them. *Note:* These implications often seem counterintuitive.

TIP

The main misperception that you need to overcome in studying quantum physics is, to put it bluntly, that you actually understand how the universe operates.

REMEMBER

In quantum physics, when your hand rests on a table, you aren't looking at two solid physical objects. You are looking at two fields of particles interacting together in a particular way. Both are made up mostly of vast, empty space, but they somehow still push against each other. Thinking of this situation as two solid surfaces pressing against each other isn't wrong, it just doesn't represent the activity going on at the quantum mechanical level. The solid surfaces that you see and feel are an outcome of all of the more fundamental quantum physical interactions.

A Matter of Scale, or a Scale of Matter

Part of the reason physicists took so long to figure out these quantum elements of physical reality is that the elements become apparent only at extremely small scales. In their normal lives, people go around interacting with large, macroscopic systems.

It's worth noting what the word *large* means in this context. A grain of sand is estimated to contain anywhere from 1 quintillion to 100 quintillion atoms. That estimate translates to more than 100,000,000,000,000,000,000 (10^{20}) atoms in a single grain of sand — or about as many atoms in a grain of sand as there are stars in the universe. And a grain of sand is so large that you don't need to take quantum physics into account to figure out how it behaves.

These large systems (relatively speaking) have the quantum effects washed out. So, although each individual atom involves quantum uncertainty, when you look at the full 100 quintillion atoms in the grain of sand, all of those quantum uncertainties cancel each other out. When viewed as a whole, the quantum uncertainty that remains on the grain of sand is completely irrelevant.

REMEMBER

Physicists began to notice quantum physics only after they could look at a single atom or, even more precisely, after they began to look *inside* of a single atom — for example, when they examined electrons within an atom or the structure of the atomic nucleus. This tiny level was where quantum behaviors really became evident.

Quantum physics discoveries enabled physicists to finally begin to understand what was going on inside of atoms. The modern understanding of atomic structures is entirely built upon the understanding of quantum physics, even though the macroscopic physical structures that come out of those atoms — whether a grain of sand, a squirrel, or a planet — don't exhibit the same quantum behaviors that you can see when looking at their smallest pieces.

Measurements and Observables: How Scientists Know Quantum Physics Is True

Quantum mechanics is a means of carefully analyzing a quantum physical situation and describing the observable outcomes of measuring a quantum mechanical experiment. Although some key insights guide the field, quantum mechanics is driven largely by the fact that the equations that are used work. The equations are complex and messy; they required years of mathematical and physical study to fully understand them. But when you do figure out how these equations work, they give you information that matches with the observable output of an experiment.

Because quantum physics relies on behavior that people don't experience in their day-to-day world, one of the biggest challenges for those studying quantum physics is learning how to rely on abstract understandings that are inconsistent with their natural intuitions.

Studying classical physics is a cakewalk by comparison because people have natural intuitions that are completely consistent with classical physics. Many a child can toss a ball at the right angle to be caught by another child. (Not me, necessarily, but many other children could do it.)

Doing the right tests

Researchers know that, to show whether anything in science is true, the explanation must match with experimental outcomes. In many cases, the findings of experimental results are fairly consistent with your intuitions, but in some cases, accepting the findings involves realizing when your intuitions are wrong.

TIP

Almost everything learned in quantum physics experimentation involves people realizing that their intuitions are wrong.

To use a historical example, thinkers going back to before the ancient Greeks believed that heavier objects fell at faster rates than lighter objects. This is a very intuitive thing to believe and is probably still the guess most children would make about how things fall.

Not only is this idea intuitive, but it is even supported by simple experiments. If you have two balls of exactly the same size, but one is made of lead and one is made of foam, the lead ball (when dropped in a simple experiment) is going to hit the ground first. But if you create a more careful experiment, which eliminates the possible impact of wind (or air) resistance, then you discover the same thing that scientists (or *natural philosophers*) discovered centuries ago: Both the balls will fall at precisely the same rate. Any experimental difference in their falling rate is actually independent of their weight. (You can find out a bit more about the laws of motion that come out of this in Chapter 2.)

REMEMBER

It wasn't enough to just do an experiment; scientists had to figure out how to do the *right* experiment. In some cases throughout history, you see that the right experiments come along with the invention of technology, such as the microscope or the telescope, which allow for tests and experiments that couldn't be done before with the naked eye.

Similarly, the technology of the industrial revolution meant that the dawn of the twentieth century brought with it the technology needed to conduct the right kinds of experiments — many of them covered in Chapter 3 — that led scientists to uncover the quantum physical nature of reality.

Trusting (but verifying) the right tests

When the results of an experiment conflict with scientists' intuitions, they realize that their intuitions are wrong — but maybe not instantly. Science has no shortage of scientists jumping to a belief that later proved to be wrong.

As the famed theoretical physicist Richard P. Feynman (whom I discuss in Chapters 3, 5, and 16) once said:

The first principle is that you must not fool yourself and you are the easiest person to fool.

Healthy skepticism is fundamental to science. Ideally, scientists carefully confirm their own results multiple times before sharing them. But, if the scientist has jumped ahead and perhaps violated Feynman's first principle, then the skepticism of the scientific community should be at the ready. Any scientist wants to be up to the task of proving a popular scientific claim wrong. Doing so is a little easier than coming up with a whole new claim (and almost as much fun).

As you find out about the history of quantum physics, notably throughout Chapters 3 and 6, you see conflicts between scientists about the proper understanding of the results of physics experiments. And, even when physicists agree on the basic concepts of how to solve quantum physics problems, they have even deeper disagreements on how to interpret those solutions!

REMEMBER

For over the last century, though, the results of quantum physics have been put to the test over and over again, and the results have constantly proven to be consistent with the mathematical solutions of the theory. Even while there is strong disagreement about the interpretation and meaning of those solutions, physicists set aside those disagreements and go on to work on applying quantum physics to build particle accelerators, create exotic superfluid states of matter, build semiconductors and transistors, and continually expand the limits of what people can create.

ENTANGLEMENT AND NONLOCALITY

One important quantum physics concept is called *entanglement*. It means that particles have their quantum states bound together due to a previous interaction between those particles.

A classic example of this is easy: Suppose that I am holding two pennies, one minted in 1967 and one minted in 1999. Without looking at either of the two pennies, I give you one of them. This system can have only two possible states:

- You have the 1967 penny, and I have the 1999 penny.
- You have the 1999 penny, and I have the 1967 penny.

These pennies are entangled. Once someone takes a measurement (looks at) one of the pennies, everyone can instantly know what state the other penny is in, even though no measurement was made of it.

Now in classical terms, this is fairly obvious. The problem is that in quantum physics (at least in the most common interpretation), the two states are fundamentally uncertain until the measurement is made. This means that there's an inherent *nonlocality* in quantum physics, meaning that two particles that are far apart from each other can be linked together in a way that doesn't make sense using the classical conception of logic. More specifically, it seems as if the particles can communicate faster than the speed of light.

I cover these concepts in more detail in Chapter 6, because they are crucial elements that quantum physics interpretations try to explain, and that criticisms of quantum physics have tried to exploit.

Chapter **2**

Standing on the Shoulders of Giants: Classical Physics

Throughout most of this book, I take you on a journey through the mysterious quantum realm — a world of microscopic objects and elementary particles — where the normal rules governing matter and energy that humans *think* they perceive in our universe don't apply. But to understand the significance of what you see during that exploration, you first need to understand the matter-and-energy rules that humans are used to seeing. That is, you need to understand some basics of classical physics.

The term *classical physics* essentially refers to any concepts and theories of physical systems that were popularly accepted from the Enlightenment era (starting roughly in the late 1600s) through to the dawn of the twentieth century. Or, to put it another way, classical physics generally covers everything from what Isaac Newton knew when he died up until what Albert Einstein learned before he began working on his doctorate (at which point he sort of revolutionized things). If you took a high school physics course, then the vast majority of what you covered was probably classical physics.

In this chapter, I explore some key ideas of classical physics and lay the groundwork for expanding these same ideas into the quantum realm. Specifically, I look into the behavior of matter when it moves and is acted on by forces. Then I cover concepts about the mathematics and physics of waves, including how this relates to a centuries-long debate regarding the fundamental nature of light. After explaining the pre-quantum resolution to that debate, I briefly explore what classical physics revealed about the existence of atoms and how energy was transferred. And I conclude by discussing probability and uncertainty in classical physics; these mathematical concepts are central to an understanding of quantum physics.

Objects in Motion: Classical Mechanics

The origins of physics as a science grew from people trying to understand the movement of objects. Objects rolling on the ground. Objects hurled through the air. Objects falling. Celestial objects moving through the heavens.

You can find some very helpful thinkers along the way — notably Nicolaus Copernicus (1473–1543), Johannes Kepler (1571–1630), and Galileo Galilei (1564–1642) — but the particulars of their discoveries culminated in the work of Sir Isaac Newton and the laws of motion that he described. Newton's work laid the foundation of the physics of motion, or *classical mechanics*.

Newton makes the rules

The accomplishments of Isaac Newton (1643–1727) include defining the laws of universal gravitation and the basic rules of optics, along with the invention of calculus. And then he also created the three laws of motion, which is what I will focus on for an introduction to classical mechanics.

Newton laid out his discoveries fairly comprehensively in his 1687 masterpiece *Philosophiae Naturalis Principia Mathematica*, or *The Mathematical Principles of Natural Philosophy*. This book contained Newton's three major contributions to classical mechanics:

>> Law of universal gravitation

>> Laws of motion

>> Calculus

Describing motion

Much of classical mechanics involves the three laws of motion in some way. The laws of motion are three statements about the relationship between an object's motion and the force acting on the object.

>> **The first law: The Law of Inertia** states that an object's state of motion will change only if the object is acted upon by a force. In a force-free situation, an object's state of motion is moving in a straight line at a constant speed (or at rest if the object's speed is zero).

>> **The second law:** $F = ma$ is a formal mathematical statement about how the force (F) on an object translates into acceleration (a) on that object. The m in the equation stands for the mass of the object.

>> **The third law: The Law of Action and Reaction** states that the force acting on an object always works in both directions. No physical body affects an object without also being affected by that object.

The initial portion of an introductory physics course is largely devoted to figuring out how to define the force or the acceleration in a given situation, or to figure out how the various forces (including the reaction forces from the third law) connect to each other.

TIP

Newton's first law really describes a special case of his second law. If no force acts on the object, then $F = 0$, meaning that $a = 0$ anytime there is no force. And if there's no acceleration at all, that means the object without a force is either remaining at rest or moving in a straight line. Similarly, if $a = 0$, then the equation says that $F = 0$.

Applying calculus to the motion

Using his laws of motion, Newton then applied his calculus to determine the forces at work on moving objects. This formed the basis of his law of universal gravitation, or law of gravity. He also confirmed that his explanation and mathematics matched up with the elliptical orbits of the planets described by Kepler years earlier.

TIP

Both the force and the acceleration are *vectors*, meaning that they consist not only of a magnitude, but also of a direction. Mass, on the other hand, is a *scalar*, meaning that it has only a magnitude, but no direction associated with it.

REMEMBER

And guess what? The unit that physicists use for force is called — wait for it — the *newton*! It is represented by the letter N, and one N equals $1 \ kg \cdot m/s^2$. If you applied the second law to a mass of 1 kg that was accelerated exactly 1 m/s², then the force involved would be exactly 1 N.

Kinetic energy and momentum

In addition to the force acting on an object, two other key concepts that apply in classical mechanics are the energy and momentum in the system associated with the object.

Energy is the capacity for doing work. This capacity can manifest in a variety of ways, but the one I focus on first is the energy of motion, or *kinetic energy*. This energy manifests when an object, with mass m, is moving at a velocity, v. Using just the mass and velocity, you can calculate the kinetic energy, KE, of the object by using this equation:

$$KE = \frac{1}{2}mv^2$$

The unit for energy is in joules (J), in the equation, $1\,J = 1\,kg \cdot m^2/s^2$. Even though the object is moving in a direction, the kinetic energy is a scalar (magnitude only) quantity. Energy is always scalar, and never has a direction.

Finding momentum when direction matters

If the direction matters, you look to the *momentum* instead of the energy. The variable for momentum is p, which comes from the term *impetus* that was used for this concept in the pre-Newton days. Because the variable m is already being used for mass, using m for both variables would be confusing, so physicists continue to use p for momentum. You will notice that momentum, p, is also calculated directly from only the mass, m, and velocity, v, by using this equation:

$$\mathbf{p} = m\mathbf{v}$$

Both the $\mathbf{p}$ and $\mathbf{v}$ are boldfaced in the equation to signify that they are vectors, with a magnitude and direction. When dealing with a vector quantity, just knowing the magnitude by itself isn't enough information. The direction, or orientation, of the vector is also important to know. Two vectors can have the same magnitude but are distinct from each other if they point in different directions. Figure 2-1 shows an example with two velocity vectors, v_1 and v_2.

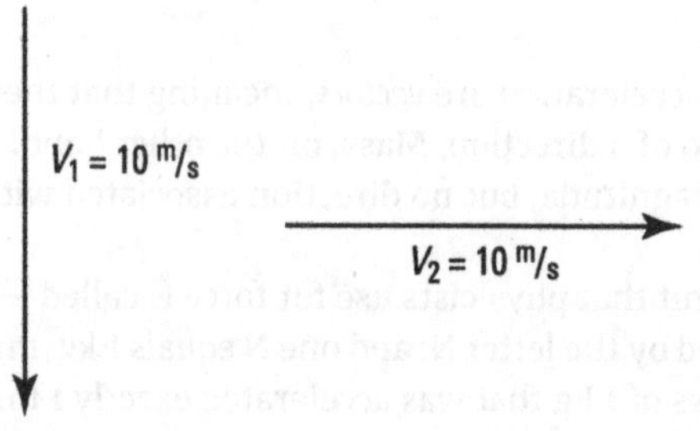

FIGURE 2-1:
Two vectors that have the same magnitude, but different directions.

$V_1 = 10\,^m/_s$

$V_2 = 10\,^m/_s$

However, the mass, *m*, in that equation is a scalar. Multiplying a vector times a scalar won't ever change the direction of the resulting vector product. In other words, the momentum, *p*, will always be in exactly the same direction as the velocity, *v*. So, if a 10-kilogram object is moving at 5 meters per second directly to the right, the momentum will be 50 kg·m/s directly to the right.

TIP

Momentum (unlike force or energy) does not have its own unit. It is measured in complex units of kg·m/s. If physicists had strongly needed a momentum unit, they likely would have defined one at some point, but it isn't really necessary. In real-world physics problems, physicists rarely seek momentum as an end goal, but instead use momentum as an intermediate step to determine the outcome related to some other unknown variable, such as force, velocity, acceleration, energy, or mass.

Focusing on kinetic energy — but others exist

Since both kinetic energy and momentum are directly calculated from mass and velocity, you may have noticed the similarity between those equations. Actually, you can eliminate the need for velocity entirely, and write the kinetic energy equation so that it depends only on the momentum and mass, in the form,

$$KE = \frac{p^2}{2m}$$

Vectors are important in physics, including in quantum physics, so it's important to mention again that momentum is a vector while energy is a scalar. But in the kinetic energy equation, when you multiply momentum times itself (square it), the directional aspect goes away in that process, and you end up with just a scalar value.

TECHNICAL STUFF

If you're a fan of calculus, you may have recognized that momentum (*mv*) is the derivative of the kinetic energy ($\frac{1}{2}mv^2$) with respect to velocity. Newton invented calculus because it was convenient for working in these sorts of situations, although you could also derive the kinetic energy and momentum equations without using calculus.

I focus this section solely on kinetic energy (the energy of motion), but here are just a handful of the other forms of energy.

>> **Gravitational potential energy:** Energy that an object has because its position changes in a gravitational field

>> **Thermal/heat energy:** Energy within a system that is responsible for its temperature

>> **Electrical energy:** Energy related to forces on electrically charged particles and their movement

>> **Chemical energy:** Energy that's stored in the bonds of atoms and molecules

>> **Nuclear energy:** Energy contained within the atomic nucleus

Catching the Waves

One way that energy manifests through movement is in waves. Most people outside of physics think of waves most commonly as the way fluids move, because they connect waves with going to the beach. But physicists find that this simple, repetitive motion (a wave) is one of the concepts that shows up most repeatedly throughout natural systems. You can find waves in the movement of fluids, of course, but you also discover waves when considering the movement of sound.

The motion of weights hanging on pendulums or springs display repetitive motion around an equilibrium point in regular oscillations, making them perfect to describe via classical waves. Chapter 9 focuses on the quantum physics approach to oscillators.

Some wavefront properties

Waves have some common characteristics that apply in every circumstance, regardless of the form of wave that you're talking about. These are most easily demonstrated in a mechanical wave like the one shown in Figure 2-2.

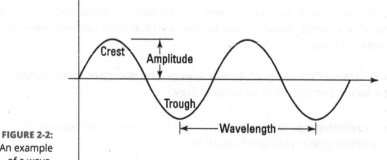

FIGURE 2-2:
An example
of a wave.

The common characteristics of waves include

>> **Amplitude (A):** The amplitude is the wave's maximum displacement from the equilibrium. The nature of the displacement can appear with

- *Constant amplitude* (refer to Figure 2-2). The resulting *continuous waves* have a displacement that is the same on every cycle.

- *Modulated amplitude*, where the displacement varies with time and/or position. On an old-fashioned radio dial, the AM stands for "amplitude modulation," because this is how the radio signal is encoded with information. Amplitude is measured in distance, in units that make sense based on the size of the wave.

>> **Crest:** The top of the wave. The height of the crest, in a mechanical wave, equals the equilibrium plus the amplitude.

>> **Trough:** The bottom of the wave. The depth of the trough, in a mechanical wave, equals the equilibrium minus the amplitude.

>> **Period (T):** The time interval required to complete one wave, cycle, or vibration.

>> **Frequency (f):** The number of waves, cycles, or vibrations that happen in a unit of time. Frequency is the reciprocal of the period.

>> **Wavelength (λ):** The distance between two successive identical points on the wave. This distance is usually measured from crest to crest or trough to trough.

>> **Speed (s):** The distance travelled by a particular point on the wave in a given interval of time. The speed equals the wavelength divided by the period $\left(s = \dfrac{\lambda}{T}\right)$ or the wavelength times the frequency ($s = \lambda f$).

You can find the two simplest mathematical waves, related to the trigonometric functions sine and cosine, shown graphically in Figure 2-3. In particular, the sine wave is useful because when the input value is zero, the output value is zero. One convenient tool in physics (including in quantum physics) is defining, or redefining, your variables so that you can apply knowledge about these trigonometric functions to help solve the problem.

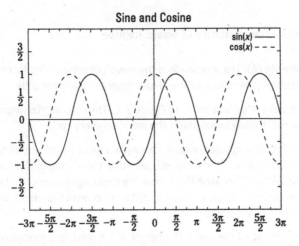

Sine and Cosine

sin(x) ——
cos(x) ----

FIGURE 2-3:
A graph of the
sine and cosine
functions.

In general, when talking about classical waves, you have two different physical types, which you find depicted in Figure 2-4.

>> **Transverse wave:** A wave that vibrates perpendicularly to the direction of propagation. For example, a water wave vibrates up and down, while the wave is moving horizontally along the surface of the water.

>> **Longitudinal wave:** A wave that vibrates in the direction of propagation. For example, with sound, the air molecules are vibrating in the same direction that the sound is moving, away from the source of the sound.

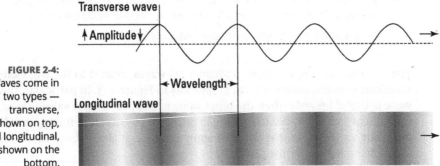

FIGURE 2-4:
Waves come in
two types —
transverse,
shown on top,
and longitudinal,
shown on the
bottom.

Transverse wave

Amplitude

Wavelength

Longitudinal wave

Whether talking about the harmonics of sound waves or the periodic motion of a mass on a spring (which I cover briefly in Chapter 9), traditional physicists use wave characteristics to model the physical behavior of these systems. The mathematics of waves become central to quantum physics, as well.

Wave interference and superposition

When dealing with waves, you must have a firm understanding of how two different waves interact. In short, the two waves overlap each other in a way that forms a *superposition* of the two waves. Instead of talking about the two distinct waves, the superposition now lets you talk about a single wave.

For example, consider the physical waves on the surface of water. If you see two sources of waves mixing together, the colliding waves don't remain as a series of distinct waves. The two wavefronts become a collective undulation of water, where the two wave sources result in a collection of waves that can no longer be distinguished from each other. You are looking at the superposition of the two waves.

The result of the superposition is *interference*. Consider two physical water waves that are exactly equal in intensity and overlap with each other. If one wave were high (at a crest) and the other wave were low (at a trough) when colliding, the resulting superposition of the wave at that point would be completely flat. The crest and the trough would cancel each other out.

The process in which the values of the two waves are added together is called *interference* and has two extreme instances:

>> **Destructive interference** (as in the preceding example) happens when a crest and trough of equal intensity cancel each other and negate the wave.

>> **Constructive interference** happens when two waves overlap so that both of their crests are coinciding. This creates a new wave crest that equals the sum of the two crests and is therefore higher than either individual wave.

Most interference doesn't line up crests and troughs perfectly, of course, and falls at some stage between destructive and constructive interference, as shown in Figure 2-5.

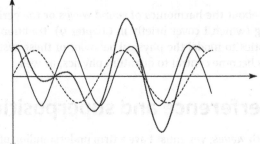

FIGURE 2-5:
When two waves overlap, the total displacement is the sum of the two individual displacements.

—— Wave #1

----- Wave #2

······· Addition of Wave #1 and Wave #2

Let There Be Light: Electromagnetism

In the time between Newton and Einstein, one of the biggest physics debates involved the nature of light. This classical debate is central to the understanding of quantum physics.

A light dispute: Corpuscles versus vibrations

Prior to Newton's work, the dominant theory interpreted light as a vibration of some unknown medium. In other words, light was interpreted as a wave-based theory, most notably by Dutch physicist Christiaan Huygens (1629–1695) and French philosopher René Descartes (1596–1650).

REMEMBER

These wave-based light theories required some sort of medium that was doing the vibrating. If light is like a wave moving through the water, some substance must act like water, even in a vacuum. Physicists came to refer to this unknown medium for light as *ether*, although sometimes they (more dramatically) called it the *luminiferous aether*.

Dealing with obstructions

To consider why thinking in terms of waves helps, you must recognize that when you have a source of light, the light seems to emanate out in all directions from that source (unless the light is blocked). This motion is like a wave rippling out in all directions from a point source, for example, if you dip your finger into the surface of a still puddle.

But, more importantly, think about what happens when an obstruction gets in the path of light. If waves in water come upon a fallen log blocking their path, when the waves reach the end of the log, they don't just keep moving in a straight line. Instead, they ripple around the log on the other side.

Something similar happens when you have obstructed light. If you have a light on in a hallway, the light moves down the hallway, but if you're in a large room at the end of the hallway, the light doesn't only illuminate whatever lies in a straight line. Some of the light expands to illuminate objects in the room that aren't in the direct path of the hallway. The light goes through the doorway and appears to bend out around the doorway, as if it's trying to illuminate as much space as possible.

This process where light encounters an obstacle and expands to try to fill the area around the obstacle is called *diffraction*. Figure 2-6 shows light waves falling on a piece of metal that has a small slit in it. The slit acts as a new source of light, with light waves radiating outward from the slit on the other side of the piece of metal (the obstruction).

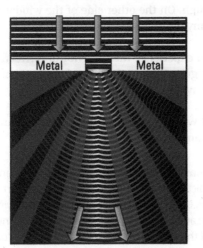

FIGURE 2-6:
Diffraction in action through a single slit.

Theories to explain light's behavior

To explain this radiant behavior, Huygens proposed that light was not merely a single wave (that always moves in the same direction), but rather a wavefront that constantly propagates itself. In his theoretical model of light, every point along the wavefront actually emits new spherical waves in all directions.

REMEMBER

This wavefront theory explained the non-linear behaviors of light quite well. The problem with this theory was that light also exhibited several linear behaviors, and Huygens's model didn't explain those particularly well.

Newton's book *Opticks* (1704) stated that experiments with the reflection and refraction of light were best explained if light moved in straight lines. Because waves don't necessarily move in straight lines, the explanation he preferred involved small particles of light, called *corpuscles*, that move in a straight line at a finite speed. This *corpuscular theory of light* dominated physics for nearly a century until new experiments made physicists once again consider whether thinking of light as a wave made more sense.

Young's double slit contradiction

In 1802, the London ophthalmologist and physicist Thomas Young (1773–1829) began what has become known as the *double slit experiment*. In this experiment, he used mirrors to direct a beam of light toward a windowpane that was mostly covered so that light could not penetrate. But the windowpane did feature two narrow slits that allowed light through. On the other side of the windowpane was a dark screen, which would show any light that shone upon it.

REMEMBER

If you have a single slit, you experience the behavior of diffraction (see the preceding section and refer to Figure 2-6), so Young had to set up the experiment just right — with two slits — to avoid the diffraction.

Consider two scenarios and what you might expect to see when a beam of light shines on Young's windowpane:

>> **If light were particles that moved in straight lines,** as Newton had suggested, then the prediction of what would appear on the screen was relatively straightforward. Either the light directed toward the windowpane would strike the covering and stop, or the appropriately positioned particles would go through one of the two narrow slits. In other words, Young would expect to see two narrow bands of light on the screen that corresponded to the corpuscles that happened to go through the two narrow slits.

>> **If the light were a propagating wavefront,** what Young actually observed would match its wavelike behavior. Instead of two narrow beams of light directed into two bands on the screen, each of the slits (as the origins of new spherical waves) would instead direct those two wavefronts to both strike the screen. Because of the waves' properties, what you'd actually see on the screen is the interference pattern from the two waves, which would result in a series of alternating light and dark bands, as shown in Figure 2-7. And this is exactly the sort of pattern that Young observed.

FIGURE 2-7:
Light moving
through the two
slits as a wave
would create a
series of light and
dark bands.

REMEMBER

The display observed on the screen of Young's experiment seemed completely incompatible with Newton's corpuscular theory, but it made sense using Huygens's wavefront model or a similar wave-based model of light. Despite having embraced Newton's corpuscular theory of light for nearly a century, physicists had to abandon the idea that light behaved like a particle. Newton's take on optics was still a useful model for performing calculations, but Young's experiment clearly demonstrated that light's behavior must be treated as wavelike.

In truth, though, neither model answered all of the questions for physicists. Neither the particle model nor the wave model fully captured the complexities of what happened with light. But in the nineteenth century, physicists were strongly leaning toward the wave interpretation of light, which meant that they had to contend with the unknown ether once again. Chapter 3 covers the double slit experiment in a quantum physics context.

Maxwell's marvelous equations

In addition to research into optics, the nineteenth century saw an amazing growth in applying magnetism and electricity to power the ever-accelerating industrial revolution. The work of Michael Faraday (1791–1867) and others in this area is worthy of a much more comprehensive treatment than I give it here. Within the span of just a few decades, these researchers took the forces that fueled parlor tricks with bell jars and magnets and turned them into a rigorous, highly mathematical science.

Connecting the forces

TIP

Significantly, nineteenth-century physicists realized a deep relationship between electricity and magnetism, as dual manifestations of the same underlying force: the *electromagnetic force*.

This electromagnetic force was carried by an *electromagnetic wave*, which was also the wave that carried light. Not only visible light — which Newton and Huygens studied in optics — but also other forms of non-visible light were just being discovered. (See the nearby sidebar, "The Electromagnetic Spectrum.") An electromagnetic wave is shown in Figure 2-8.

In the twentieth century, quantum physics transformed the understanding of both light and electromagnetism. But the foundation for that transformation was this amazing work in the nineteenth century, where physicists learned that deep mathematical relationships exist between phenomena that appear to be wildly different.

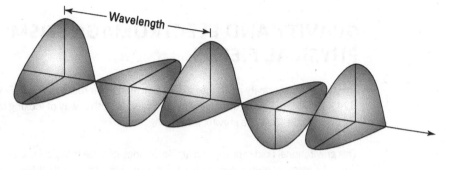

FIGURE 2-8: The electric force and magnetic force are in step in an electromagnetic wave.

Expressing the connections with math

The electromagnetic wave reached its full expression in a series of four equations that are called Maxwell's equations — published in 1865 — in honor of James Clerk Maxwell (1831–1879). In truth, he didn't create all of the equations himself (as you can tell because several of them are individually named after different people), but he's credited with pulling them together into a coherent mathematical understanding of electricity, magnetism, and light in the form of electromagnetic waves.

Although relatively short, these equations contain a surprisingly complex amount of information. Among the information that Maxwell would extract from the equations is the realization that the speed of the electromagnetic field precisely matches the speed of light, which led him to understand the deep and profound relationship between these phenomena.

Here are a couple of points to help you interpret Maxwell's equations:

TECHNICAL STUFF

>> **The boldface letters** reference the electric field (**E**) or the magnetic field (**B**), which are represented by vector fields. (See the nearby sidebar, "Gravity and Electromagnetism: Physical Fields.")

>> **The del or nabla symbol,** ∇ (which appears in the Chapter 7 discussion of vector operators), can represent the following three aspects of a vector field, **F**, when used in vector calculus.

- $\nabla \mathbf{F}$: Gradient of **F**

- $\nabla \cdot \mathbf{F}$: Divergence of **F**

- $\nabla \times \mathbf{F}$: Curl of **F**

You can represent Maxwell's equations in several ways, but Table 2-1 shows some of their more common forms as differential equations.

GRAVITY AND ELECTROMAGNETISM: PHYSICAL FIELDS

Newton defined gravity as an interaction between point masses. Pierre-Simon Laplace (1749–1827) came along after Newton and found another way of looking at gravity, by introducing the idea of a *gravitational field*.

This gravitational field represented gravity as lines of force that permeated all of space around a material object, such as Earth. At every point near Earth, there is a gravitational field with an intensity based on Earth's mass and the distance of that point from the center of Earth. This gravitational field has not only a magnitude at every point, but also a direction (down in this case, toward the center of Earth's gravity), meaning it's a *vector field*. If another mass is within that field, then the gravitational force between it and Earth is based on their two masses, in a way that precisely coincides with Newton's calculations.

The gravitational field approach became particularly popular throughout the nineteenth century. In fact, fields became a common way of thinking about physics ideas, largely because it was also a natural way of working with electromagnetic forces. Instead of a gravitational field, these forces could be interpreted as coming from an *electromagnetic field* that filled the area around any charge or current. In particular, you can even see the lines of force implied by magnetic fields around a magnet using iron filings!

TABLE 2-1   ## Maxwell's Equations

Equation	What It's Called
$\nabla \cdot \mathbf{E} = \dfrac{\rho}{\varepsilon_0}$	Gauss's law
$\nabla \cdot \mathbf{B} = 0$	Gauss's law for magnetism
$\nabla \times \mathbf{E} = -\dfrac{\partial \mathbf{B}}{\partial t}$	The Maxwell-Faraday equation or Faraday's law of induction
$\nabla \times \mathbf{B} = \mu_0\left(\mathbf{J} + \varepsilon_0 \dfrac{\partial \mathbf{E}}{\partial t}\right)$	The Ampere-Maxwell equation, an extension of Ampere's circuital law

Michelson-Morley and the mysterious missing ether

By the middle of the nineteenth century, it certainly seemed to physicists that Newton had decidedly lost the debate on the nature of light. Even Newton, apparently, couldn't be right about everything. Maxwell's equations seemed to thoroughly describe the behavior of light, as well as electricity and magnetism, as electromagnetic waves. Though Newton's optical approaches were still useful for solving problems of refraction and reflection of light, everyone informed on the topic thought that light was, indeed, a wave.

Looking for the ether

But a significant question remained: What medium was doing the waving?

The standard answer was that the medium was ether, or luminiferous aether (depending on how fancy you wanted to sound), but the problem is that this basically just meant "the unknown thing that is transmitting a light wave." Light could travel through a vacuum, after all, so if ether was the thing that was waving, then it had to be a substance that permeated through all of space. The fact that physicists weren't able to detect it at all was a stumbling block, to say the least.

REMEMBER

Although physicists didn't know anything about how the ether behaved, they did know some things about how waves behaved. If the medium is moving, then a wave moving through the medium will be affected. To return to the example of water, liquid waves moving in a river that has a flowing current move faster in the direction of the current than they move toward the edge of the river.

The assumption, though, was that ether probably didn't move. It was probably some fundamental aspect of the universe. But even if ether didn't move, physicists knew that the Earth was moving, which meant that everything on Earth was moving relative to the ether.

To try to detect the ether's movement with this knowledge, physicists built a device called an *interferometer*. You can find a few different designs for these devices, and Figure 2-9 shows a mock-up of the Michelson–Morley interferometer, which adequately depicts what I describe in this section.

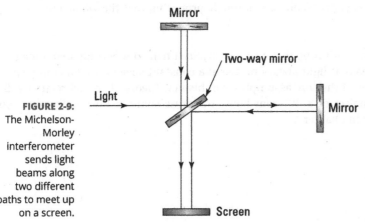

FIGURE 2-9: The Michelson–Morley interferometer sends light beams along two different paths to meet up on a screen.

Comparing the light on two paths

So, what is going on with this contraption? Light enters the device, and then special mirrors allow some of the light to pass through in a straight line, and some of the light is then deflected in a perpendicular direction. Both of those beams of light strike other mirrors and are sent back, bouncing off the special mirror so that they are both heading toward a screen. Scientists applied some parameters and assumptions when experimenting with an interferometer:

>> **The distances between the mirrors are identical,** so the two light beams have travelled exactly the same distance on each path to hit the screen. If nothing else were affecting the speed of the light, then the light waves *should* be in phase with each other. This means that you'd see a bright spot of light on the screen.

>> **The light beams may have travelled exactly the same distance,** but one of the paths would move either with the ether or against the ether. In that case, the light that went down that path would move either slightly faster or slightly slower than the light that went along the other path. When the two paths of light rejoin, then, the ether theory predicts that, as wavefronts, they would be slightly out of phase with each other, resulting in an interference pattern of light and dark bands on the screen.

REMEMBER

Physicists Albert Michelson (1852–1931) and Edward Morley (1838–1923) conducted a series of experiments hoping to find the effect of ether in 1887, but their results were disastrous (for the ether theory, at least). None of their attempts detected different speeds along the different paths of the light. *It was almost like the ether didn't actually exist!*

But the evidence in support of the wave theory was overwhelming at that point, so most physicists (including Michelson and Morley themselves) just assumed it was a failed experiment. No one was strongly proposing that the whole ether idea be thrown out.

It wouldn't be until 1905 that a young physicist named Albert Einstein took seriously the idea that light always moved at a constant speed through empty space, making it one of the core assumptions in his new theory of special relativity. But that's the subject of a different book. I address some of Einstein's other 1905 observations in Chapter 3.

Atoms: Building Blocks of Matter

In his book of physics lectures, *Six Easy Pieces*, famed physicist Richard Phillips Feynman (1918–1988) depicts a hypothetical catastrophe that destroys all human knowledge. In this scenario, he has the ability to pass along only one sentence to future generations. What should it be? What tidbit can be put forward to contain the most knowledge in the shortest number of words? Feynman put forward this *atomic hypothesis* as his proposal:

> "All things are made of atoms — little particles that move around in perpetual motion, attracting each other when they are a little distance apart, but repelling upon being squeezed into one another."

You find out more about the curious character of Feynman and his contributions to the field of physics, in Chapters 3, 5, and 16. By the time he made this statement, he certainly knew an amazing amount about the quantum structure of the atom.

TIP

What is amazing about Feynman's atomic hypothesis is that the first part of this sentence, "All things are made of atoms," is one of science's most ancient ideas but also surprisingly recent.

Ancient atomism

The idea of the atom dates back to the ancient Greeks, most notably to Democritus. The basic argument for the existence of atoms came down to this: It is impossible to keep dividing something forever. At some point, a final tiny bit of *something* that cannot be broken down any further must exist. That thing, said the ancient atomists, is the atom.

The idea came and went. Epicurus (341–270 BC) and then later Lucretius were both fans of the idea of atoms. But even before a full scientific method of investigating explanations was in place, the problem with this idea was that no one could ever get to the point where they reached the atom. No matter how small something was, you could always break it down further. Plato spoke in favor of atoms, but Aristotle was opposed, and Aristotle tended to carry the day.

The idea of the atom would reoccur occasionally throughout the intervening millenia, including atomic theory variants put forth by Galileo Galilei, René Descartes, and Isaac Newton, but it wasn't until scientists came up with evidence that the idea of atoms became widely adopted.

Chemists discover the atom (maybe)

Chemistry developed throughout the 1700s and recognized that some substances were elements that couldn't be broken down any further. Other substances were compounds, made from combining multiple elements together. Water, for example, came from combining the right ratio of hydrogen and oxygen.

British chemist John Dalton (1766–1844) compiled the chemists' work together in 1808, noting that the ratios of elements to the resulting compounds was always in whole numbers. Nothing in the data collected suggested that you could take half of an oxygen element and combine it with one hydrogen element to get a molecule from that, for example.

The studied elements all gave the impression that they couldn't be broken down any further, so Dalton put forward the idea that at their smallest structural form, these elements were atoms. Basically, Dalton recognized that chemistry is about separating atomic elements from each other and joining them together in different combinations.

This claim triggered a major debate among chemists throughout the 1800s. Although chemists spoke about elements, not everyone embraced the idea that they actually couldn't break down any further. It is hard today to even conceive of how you would interpret the periodic table of the elements without a deep understanding of atomic structure, but even its creator, Dmitri Mendeleev (1834–1907), expressed disbelief about the idea that atoms were actual physical objects instead of just useful mathematical tools.

REMEMBER

In 1905, a certain wild-haired German physicist working in a Swiss patent office sat down to write his doctoral thesis in which he provided a detailed atom-based explanation for the well-known, but unexplained, phenomenon of Brownian motion, as explained in Chapter 3. When his predictions were experimentally confirmed in 1908, the adoption of the atomic model would really take off in earnest. If you haven't guessed, the German physicist's name was Albert Einstein, who had a habit of taking useful mathematical ideas seriously and coming up with great insights.

And the electron

One of the last pre-quantum discoveries about the fundamental nature of matter was the 1897 discovery of the electron by J.J. Thomson (1856–1940). It's amazing to think that all the discoveries about electricity throughout the nineteenth century came about without physicists even being able to detect that electrons existed.

SPECTROSCOPY: STUDYING ELEMENTS WITH LIGHT

The study of light and matter began to coincide in the mid-1800s with the development of *spectroscopy*, a branch of science related to the study of the electromagnetic spectrum emitted by matter. While physicists had long been interested anytime matter began emitting light, the tools of optics became refined enough in the middle of the nineteenth century for physicists to become more systematic in their study. This tool development, combined with the greater understanding of the electromagnetic spectrum codified in Maxwell's equations, meant that spectroscopy became a rigorous field of study.

Key in relating this to the quantum physics story were the German physicists Robert Bunsen (1811–1899) and Gustav Kirchhoff (1824–1887). In the 1860s, these two physicists began publishing work that systematically laid out the connection between certain spectral emission lines and the elements that emitted those lines. To put it simply, if you looked very carefully at the light emitted when you burned an element, you could determine which element it was. It was the analysis of these spectra that would provide much of the experimental progress in quantum physics in the next century, particularly in understanding the structure of the atom.

Spectroscopy set the stage not only for much of the quantum research that I explain in Chapter 3, but also for a transformation in our understanding of the night sky (and the daytime sky, for that matter). Within less than a decade, the married astronomers William and Margaret Huggins would use the methods of spectroscopy outside the laboratory to look at the light emitted from stars. By doing so, they were able to show that stars are made of the same elements found on Earth.

Physicists had, however, hypothesized that electrons existed. In fact, Benjamin Franklin (1706–1790) himself put forth an idea that charged particles moved around within electrical apparatuses. You may have heard the story involving Mr. Franklin and a kite, which was an example of how interested he was in this subject.

Prior to Thomson's discovery, George Johnstone Stoney used the term *electron* in 1891 to describe the unit of charge in experiments that passed electrical current through chemicals. But, still, the technology to actually detect these charged particles didn't exist until the late 1800s, with the invention of the cathode ray tube, shown in Figure 2-10.

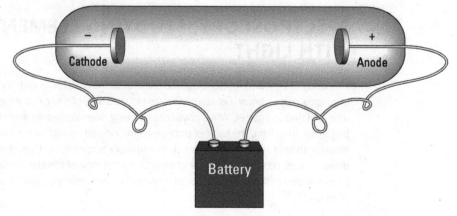

FIGURE 2-10:
Cathode ray
tubes allow
charged particles
to be studied
in a vacuum.

The tube shown in the figure contains a vacuum, so all air has been pumped out of it. There is a metal plate at each end, and those plates are connected to the two terminals of a battery. The metal plate that is positively charged is called the *anode*. The metal plate that is negatively charged is called the *cathode* (and gives the device its name).

When the power is connected and switched on, the tube begins to glow. Thomson's work showed that the visible beam was the result of negatively charged particles flowing across the tube. Thomson also went on to suggest that these particles were part of the atoms that got knocked free, somehow, from the current flowing through the device.

So it was that physicists discovered the electron, which is a lighter particle that can be knocked loose from the atom, years before they actually even confirmed for sure that atoms exist, let alone what their internal structure looked like. Figuring out the relationship of the electron to the rest of the atomic structure would be one of the great early achievements of the young quantum theory.

Thermodynamics: Another Hot Topic

An outcome of the industrial revolution was a growing body of research on translating energy into productive work and, in particular, how that energy could be efficiently produced and transferred to where it was needed. The main form of energy during the industrial revolution was heat, and the study of how heat relates to other forms of energy is called *thermodynamics*.

Earlier in this chapter (in the section, "Objects in Motion: Classical Mechanics"), I discuss energy within mechanics as focusing on the motion of a specific object.

In the field of thermodynamics, you don't care about the motion of specific particles or objects but, instead, are concerned with the internal state of the object or the system as a whole. Scientists now know that the internal state of heat or temperature is caused by the vibration of particles within the system, but (as I discuss in the previous section) chemists and physicists didn't even know that atoms existed when they developed the field of thermodynamics!

By the 1860s, scientists had developed two core *laws of thermodynamics* that governed their thinking about it. And as the study of the atom progressed (after the discovery of quantum physics), you got a third law:

>> **First law:** Energy cannot be created or destroyed.

>> **Second law:** For a spontaneous process, the entropy (disorder) of the universe increases.

>> **Third law:** The entropy of a system at absolute zero (temperature) is a well-defined constant.

The study of efficient heat systems ultimately led the German government to put out a call for researchers to help create a table of thermodynamic values for something called a *black body*. In Chapter 3, you find out how a German physicist solved the black-body problem and set in motion the first, central idea of quantum physics. (No, his name wasn't Albert Einstein. It was Max Planck. C'mon. . .Einstein couldn't be expected to do everything!)

The Games People Play: Unknowns and Uncertainties

Classical physics built its vision of the world on the idea that everything was ultimately very well-determined. If you knew all of the forces involved in all of the pieces, then you could figure out the outcome of any set of interactions. The second law of thermodynamics (as mentioned in the previous section) put an upper limit on the capacity to know all the forces and pieces, because some disorder is introduced in any spontaneous process. You can never perfectly figure out every outcome. But overall, the classical world was still highly deterministic — at least in principle.

Even in a fully deterministic universe, though, all sorts of things are impossible to determine, practically speaking. Mathematics gives you ways to deal with the uncertainty that shows up even in science before quantum physics comes along.

A roll of the dice: Classical probability

One classic example of a random event with a well-defined outcome is rolling a die. Technically, if you knew the precise force, velocity, and angles with which the die was released, and also knew what surfaces the die would bounce off of, calculating the outcome of rolling the die should be possible. For readers who (like me) are fans of a wide array of board games that use different shapes of die, let me stipulate that here I'm referring to the run-of-the-mill six-sided die, a cube with sides numbered 1 to 6 — a d6, as players call it in *Dungeons & Dragons*.

Figuring the probability of an outcome for one die

But so many factors affect the roll of a d6 that you really can't accurately gain enough information to predict the outcome. Determining the probability of any specific outcome means calculating a numerical description of how likely an event is to occur.

In the case of rolling a single die, if I wanted to know the probability of an outcome of 1, I would first have to figure out the *sample space*, the collection of all possible outcomes. For a single die roll, that sample space consists of six numerical values from 1 to 6. You can write the sample space mathematically as the set $\{1, 2, 3, 4, 5, 6\}$. Only one of those values is the outcome of 1 that I'm looking for, so the resulting probability is 1 out of 6, or $\frac{1}{6}$.

REMEMBER

Probability is a fractional representation of how often, within a given set of tries, you'd expect to get the proposed outcome. The lowest possible value is 0 (no chance), and the highest possible value is 1 (a certainty).

Consider these two examples regarding rolling a specific value on a single die. If I asked

>> **What is the probability of rolling a 7?** The answer is that the probability is 0. It is impossible to roll a 7 on that d6 die.

>> **What is the probability of rolling a value less than 7?** The answer is that the probability is 1. Every roll on a single d6 die is guaranteed to meet that criterion.

Letting probabilities simplify your analyses

TECHNICAL STUFF

Rolling one die is a trivial example, but representing the sample space and probability of outcomes quickly becomes more complex for systems that are more complicated. Consider rolling two dice. Now the sample space is much more complex. There are 6 possible values on the first die, and 6 possible values on the second die, which means that you have a total of 36 elements in the sample space, representing the values possible on two dice, shown in full here:

$$\begin{bmatrix} (1,\,1) & (2,\,1) & (3,\,1) & (4,\,1) & (5,\,1) & (6,\,1) \\ (1,\,2) & (2,\,2) & (3,\,2) & (4,\,2) & (5,\,2) & (6,\,2) \\ (1,\,3) & (2,\,3) & (3,\,3) & (4,\,3) & (5,\,3) & (6,\,3) \\ (1,\,4) & (2,\,4) & (3,\,4) & (4,\,4) & (5,\,4) & (6,\,4) \\ (1,\,5) & (2,\,5) & (3,\,5) & (4,\,5) & (5,\,5) & (6,\,5) \\ (1,\,6) & (2,\,6) & (3,\,6) & (4,\,6) & (5,\,6) & (6,\,6) \end{bmatrix}$$

If you were trying to figure out the probability of rolling 1 on both dice, then the only element that would match that outcome is (1, 1). That's one element out of 36, so it's a probability of 1 out of 36, or $\frac{1}{36}$.

But suppose that, instead, you were trying to find the probability of rolling a sum of exactly 7 on both dice. In this case, the elements (1, 6), (2, 5), (3, 4), (4, 3), (5, 2), and (6, 1) all fit this case, so the probability of rolling a total of 7 is 6 out of 36, or $\frac{6}{36}$, which reduces down to $\frac{1}{6}$. This is the most common sum of rolling two dice.

The point of working with the classical probabilities is that it's way easier to work with the fraction $\frac{1}{6}$ than to work with the full sample space, having to go back to look across 36 elements every time you want to analyze anything about the situation. You condense the detailed information about the sample space down into probability. You can then move forward working with the probabilities. In quantum physics, probability takes on a fundamentally different meaning, as you discover in Chapter 3, but it continues to have this benefit of condensing detail.

Uncertainties and deviations

In the preceding section, I talk about probability in a case where I would know the result to be a very definite outcome (rolling a die). Assuming the die is well constructed, no ambiguity about the result exists. I can just read the value showing on the face-up side of the six-sided cube. I do have uncertainty about what the actual value will be before I roll the die, but absolutely no uncertainty about the measurement once the die is rolled.

Unfortunately, in science, this certainty of the outcome (in the case of rolling a die) isn't always the case. Probabilities and uncertainties both play a central role in quantum physics. Over a century of experimental results confirm that the probability calculations of quantum physics hold up, even in ways that can seem counterintuitive. Fundamental uncertainty lies at the heart of one of quantum physics' core principles. (See Chapter 3 for a closer look at this principle.)

Accounting for measurement uncertainties

Lack of precision and differences in people's perception are potential sources of uncertainty in an experiment's measurements. Physics has fortunately accounted for these uncertainties. When a physicist records a measurement, they not only record the value but also keep track of how precise that measurement is. A physicist might measure to the nearest tenth of a meter and, by doing so, can say that a distance is 1.2 meters. But they don't know whether the measurement was actually 1.22 meters or 1.23 meters, because they didn't measure that precisely (to the nearest hundredth). Any subsequence calculations have limited precision based on that 1.2-meter measurement. If a physicist wants more precise results in the end, they must make extremely precise measurements at every step in the process.

Even in a very well-defined experiment, no two instances of the same experiment should be expected to have precisely the same result. For this reason, physicists usually try to conduct multiple iterations of their experiment, either at the same time or one after another, to gain a series of results. Each result has its own level of confidence based on the measurements made, of course.

REMEMBER

You can take the average of the measurements, but the average is only so helpful because you must somehow account for the variation that you observed across those experiments.

Putting statistical notions to use

The statistical notions of *variance* and *standard deviation* enter the picture to help physicists account for the differences in experimental results. You calculate the variance, σ^2, as the average of the square of spread in each measurement. The square root of the variance, σ, is the standard deviation. For various reasons (known by statisticians), standard deviation is much more commonly used than variance as a measure of spread in data.

For an outcome variable where you know the probability of its occurrence, finding the variance is relatively easy. You know the probability of each possible outcome, so you can figure out the probability of a given spread and use that to calculate the variance across all possible outcomes.

But for a *random variable* (a variable whose probability you don't know), calculating variance gets a little more complex. Because the variable's probability is unknown, you cannot multiply by that probability. You have to calculate the variance from a series of results. In that case, to calculate the variance, you have to consider the expectation value of the variable. The expected value of an unknown variable X could be written as $\langle X \rangle$. (I explain this notation in Chapter 7.)

To calculate the variance of such an unknown variable, X, you'd take the expectation value of the variable squared, $\langle X^2 \rangle$, and subtract the square of the expectation value, $\langle X \rangle^2$, giving you the equations for variance (σ^2) and standard deviation (σ):

$$\sigma^2 = \langle X^2 \rangle - \langle X \rangle^2$$

$$\sigma = \sqrt{\langle X^2 \rangle - \langle X \rangle^2}$$

Chapter **3**

The Quantum Revolution

The origins of quantum physics are rooted in a transformation in the way physicists think about particles and waves. No one was looking for this transformation. The nineteenth century ended with many physicists quite confident that — because they'd learned so much about energy and matter, and their interaction during the preceding centuries — they could simply tidy up the explanations for a few existing experiments.

No one realized that these explanations — such as the double slit experiment and black-body radiation — would usher in a fundamental change in physicists' viewpoints. Throughout the first half of the twentieth century, nearly everything these scientists thought they knew (about matter, energy, and how they interact) took on a new perspective.

In this chapter, I cover the major discoveries that defined quantum physics, from its origins through to groundbreaking work in the 1960s. I look into the first experiments that revealed the quantum nature of reality and provide information on how those experiments led to a new understanding of probability at the quantum level. Finally, I describe how the new quantum theory transformed the physical understanding of energy and matter.

Being Discrete: The Trouble with Black-Body Radiation

The first major idea that distinguishes quantum physics from classical physics is *quantization*, or measuring quantities in discrete, not continuous, units. Where did this idea first come from? It wasn't just that someone came up with this idea one day and decided to build an entire scientific field around it. The idea of quantized energies arose from one of the earliest challenges to classical physics: the problem of *black-body radiation*.

When you heat an object, even before that object visibly glows, it radiates light in the *infrared spectrum* (which is light you can't see, with a longer wavelength and lower frequency than visible light). When heated enough, the electrons on the surface of the object are agitated thermally, and electrons being accelerated and decelerated in this fashion radiate light in the visible spectrum. This effect is what's going on when you see a hot object, such as heated metal, glow.

Physics in the late-nineteenth and early-twentieth centuries was concerned with the spectrum of light being emitted by black bodies. A *black body* is a piece of material that radiates energy (in the form of light) corresponding to its temperature — but it also absorbs and reflects light from its surroundings. Physicists like to tackle complicated problems by assuming extreme cases. To make matters easier, physicists focused only on the emission of light rather than the absorption of light. In 1859, German physicist Gustav Kirchhoff postulated the existence of a black body that reflected nothing and absorbed all the light falling on it (hence the term black body, because the object would appear perfectly black since it absorbed all light falling on it). But when you heat a black body, it will radiate in accordance with its temperature rise and emit light.

REMEMBER

This black-body concept may seem purely abstract, but it did have some relevance to the growing electric power industry. The German government wanted to establish a formula that would show how the normal spectrum (the intensity and frequency) of this radiation varied with temperature. Historically, this experiment wouldn't be all that important — except that solving it introduced the idea of the quantum into physics!

Well, it was hard to depict a physical black body — after all, what material absorbs light 100 percent and doesn't reflect anything? But the physicists were clever about this depiction, and they came up with the hollow cavity with a hole in it that you can see in Figure 3-1. If you look at this black body anywhere except the hole, then the light is completely absorbed, and you don't see the black body.

If you shine light on the hole, all that light goes inside, where it is reflected off the inside of the cavity again and again — until it all gets absorbed. (A negligible amount of light escapes through the hole.) And if you heat the hollow cavity, the hole begins to glow because it is radiating light energy. So, there you have it — a pretty good approximation of a black body.

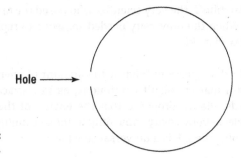

FIGURE 3-1:
A black body.

Having come up with this idea, experimenters tried to see what would happen — that is, what light spectrum would be emitted when they heated the black body. You can see the spectrum of a black body (and attempts to model that spectrum) in Figure 3-2, for two different temperatures, T_1 and T_2. The problem was that nobody was able to come up with a theoretical explanation for the spectrum of light generated by the black body. Every solution (or explanation) that classical physics could come up with failed to fit the data. The first attempt, called Wien's Law, fit for high frequencies but failed for lowers, and the second attempt, Rayleigh-Jeans Law, fit for very low frequencies but failed for higher frequencies. Researchers needed a new approach!

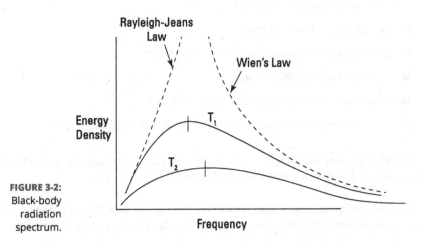

FIGURE 3-2:
Black-body
radiation
spectrum.

An intuitive (quantum) leap: Max Planck's spectrum

The black-body radiation problem was a tough one to solve, and with it came the beginnings of quantum physics. A revolutionary insight would come from German physicist Max Planck, who was in an excellent position to solve it. One of his mentors had been Gustav Kirchhoff, the very scientist who posed the black body question in the first place. When the university needed someone to replace Kirchhoff, they gave the job to Max Planck!

Here was Planck's radical suggestion: What if the amount of energy that a light wave can exchange with matter wasn't continuous, as in classical physics, but discrete? In other words, Planck proposed that the energy of the light emitted from the walls of the black-body cavity came only in integer multiples, as shown in the following equation, where h is a universal constant:

$$E = nh\upsilon, \text{ where } n = 0, 1, 2,...$$

With this theory, crazy as it sounded in the early 1900s, Planck converted the continuous integrals used by Rayleigh-Jeans to discrete sums over an infinite number of terms. Making that simple change gave Planck the following equation for the spectrum of black-body radiation:

$$u(\upsilon, T) = \frac{2\pi h \upsilon^5}{c^3 \left(e^{h\upsilon/kT} - 1 \right)}$$

Here are the pieces in this equation:

>> $u(\upsilon, T)$: The intensity distribution of the light spectrum at frequency υ of a black body at the temperature T

>> υ: Frequency of the black body

>> T: Temperature in degrees Kelvin of the black body

>> c: Speed of light, 2.998×10^8 m/s

>> k: Boltzmann's constant, representing 1.3807×10^{-23} J·K^{-1}

>> h: Planck's constant, a new constant with a value of 6.626×10^{-34} J·s

Planck's equation got it right — it exactly describes the black-body spectrum, at both low and high (and medium, for that matter) frequencies.

This idea was quite new. What Planck was saying was that the radiating *oscillators* (the mechanisms that emit the energy) in the black body couldn't take on just any level of energy, as classical physics allows; they could take on only specific, quantized energies. In fact, Planck hypothesized that that was true for any oscillator — that its energy was an integral multiple of $h\upsilon$ (the universal constant and the frequency).

REMEMBER

And so Planck's equation came to be known as Planck's quantization rule, and h became Planck's constant. Saying that the energy of all oscillators was quantized was the birth of quantum physics.

Planck's approach was not obvious, since classical physics would allow energies to take on a full range of values. Planck had applied a mathematical trick that fit the data, but that's all he thought that it was. He made no revolutionary claims about the underlying structure of light or energy. He offered no physical explanation for why the oscillators could only take on discrete energy values. Although it would take years for anyone to realize the implications of Planck's solution, the unexpected revolution had begun — and there was no stopping it.

THE QUANTUM NOBELS

The key discoveries of quantum physics led to an avalanche of Nobel Prizes in Physics. This isn't a comprehensive list, but it contains some of the key discoveries and insights that were recognized by Nobel Prizes. In particular, Nobels related to nuclear physics, rather than quantum theory itself, are largely excluded (even though there really isn't a nuclear physics without quantum theory).

- 1918 — Max Planck, for discovering the quanta

- 1921 — Albert Einstein, for explaining the photoelectric effect, by means of quantized light (photons)

- 1922 — Niels Bohr, for using quantum theory to explain the structure of the atom

- 1923 — Robert Andrews Millikan, for experimental confirmation of electron charge and the photoelectric effect

- 1926 — Jean Perrin, for experimental confirmation of atoms

- 1929 — Louis de Broglie, for the discovery of the wave nature of electrons

- 1932 — Werner Karl Heisenberg, for the creation of quantum mechanics

- 1933 — Erwin Schrödinger and Paul Dirac, for developing and expanding the mathematical approaches to quantum mechanics

- 1945 — Wolfgang Pauli, for establishing the Pauli exclusion principle governing electron behavior

(continued)

(continued)

- 1954 — Max Born, for determining that the quantum wave function represents probabilities

- 1965 — Sin-Itiro Tomonaga, Julian Schwinger, and Richard Feynman, for developing quantum electrodynamics (the quantum theory of electromagnetism)

- 1969 — Murray Gell-Mann, for discovery of the quantum theory of fundamental particles and their interactions (quantum chromodynamics)

Seeing Light as Particles

Light as particles? Isn't light made up of waves? The fundamental makeup of light had long been a topic of debate in previous centuries because evidence supported both ideas. But by the end of the eighteenth century, physicists had largely agreed that the wave interpretation of light fit most of the data, as I discuss in Chapter 2. Reality, though, was going to throw a curveball and show them that light exhibits properties of both particles and waves.

Solving the photoelectric effect

The photoelectric effect was another one of those experimental results that created a crisis for classical physics around the turn of the twentieth century. Explaining this effect was also one of Einstein's first successes, and it provided proof of the quantization of light. Here's what happened.

When you shine light onto metal, you get emitted electrons, as Figure 3-3 shows. The electrons in the metal absorb the light you shine, and if they absorb enough energy, they're able to break free of the metal's surface. According to classical physics, light is just a wave, and it can exchange any amount of energy with the metal. When you beam light on a piece of metal, the electrons in the metal should absorb the light and slowly get up enough energy to be emitted from the metal. The idea was that if you were to shine more light (that is, more overall energy) onto the metal, the electrons should be emitted with a higher kinetic energy. And very weak light shouldn't be able to emit electrons at all, except after a matter of hours.

REMEMBER

But that's not what happened in early photoelectric experiments. The electrons weren't cooperating with classical physics ideas. Instead, electrons were emitted as soon as someone shone light on the metal. In fact, no matter how weak the intensity of the incident light (and researchers tried experiments with such weak light that they expected hours to pass before they saw any electrons emitted), electrons were emitted. Immediately.

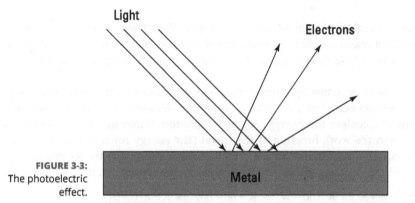

FIGURE 3-3:
The photoelectric
effect.

Experiments with the photoelectric effect showed that the kinetic energy, K, of the emitted electrons depended only on the frequency — not the intensity — of the incident light, as you can see in Figure 3-4.

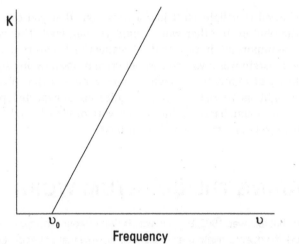

FIGURE 3-4:
Kinetic energy of
emitted electrons
versus frequency
of the incident
light.

In Figure 3-4, υ_0 is called the threshold frequency, and if you shine light with a frequency below this threshold on the metal, no electrons are emitted. The emitted electrons come from the pool of free electrons in the metal (all metals have a pool of free electrons), and you need to supply these electrons with an energy equivalent to the metal's *work function*, W, to emit the electrons from the metal's surface. In other words, the work function is the minimum energy needed for the metal to release free electrons.

The results were hard to explain classically, so enter Einstein at the beginning of his heyday, around 1905. Encouraged by Planck's success (see the section, "An intuitive [quantum] leap: Max Planck's spectrum," earlier in this chapter), Einstein postulated that not only were oscillators quantized, but so was light — into discrete units called *photons*. (He didn't call them that. The term *photons* wouldn't be coined for about another ten years and would take about another decade to be widely adopted.) Light, he suggested, acted like particles as well as waves.

So, in this scheme, when light hits a metal surface, photons hit the free electrons, and an electron completely absorbs each photon. Einstein then applied Planck's constant to calculate the energy, $h\upsilon$, of the photon. When that energy becomes greater than the work function of the metal (the energy for that threshold frequency, υ_0), the electron is emitted. The variable K represents the kinetic energy of the emitted electron, giving you this equation, which you can then solve for K:

$$h\upsilon = W+K \Rightarrow K = h\upsilon - W \Rightarrow K = h\upsilon - h\upsilon_0 \Rightarrow K = h(\upsilon - \upsilon_0)$$

Since the threshold frequency doesn't change for an object, and Planck's constant is, well, constant, this shows you that the kinetic energy of the freed electrons is entirely dependent upon the frequency of the light. And, sure enough, this approach fits the data!

REMEMBER

Einstein showed that light isn't just a wave, but that you can also view it as a particle, the photon. In other words, light is quantized. This was an important finding — so important, in fact, that it became the discovery that was specifically cited when Einstein was awarded a Nobel Prize in 1921. (Many still considered his famous theory of relativity somewhat controversial at that time.) Back in 1905, explaining light as particles was also quite an unexpected piece of work by Einstein, even though based on the earlier work of Planck. Light quantized? Light coming in discrete energy packets? What next?

SPREADING THE QUANTUM WORD

In the years that followed 1905, Einstein wasn't the only one who stumbled upon the need to work the quantum into their explanations. Now that physicists knew to look for it, they were finding multiples of $h\upsilon$ all over the place. Nothing in the microscopic world appeared to be continuous but was all quantum in nature, and this could no longer be ignored!

Belgian industrialist Ernest Solvay sponsored a conference in 1911 that brought together a group of 21 of the leading European scientists to Brussels to debate the matter. Among them was Max Planck, who claimed (about the quantum) that "for the past ten years nothing in physics has so continuously stimulated, excited, and irritated me." Planck went on to say that the existing paradigm of classical physics "is obviously too narrow to account for all these physical phenomena which are not directly accessible to our coarse senses."

Scattering light off electrons: The Compton effect

To a world that still had trouble comprehending light as particles (see the preceding section), Arthur Compton (1892–1962) supplied the final blow with the *Compton effect*. His 1922 experiment involved scattering photons off electrons, as Figure 3-5 shows.

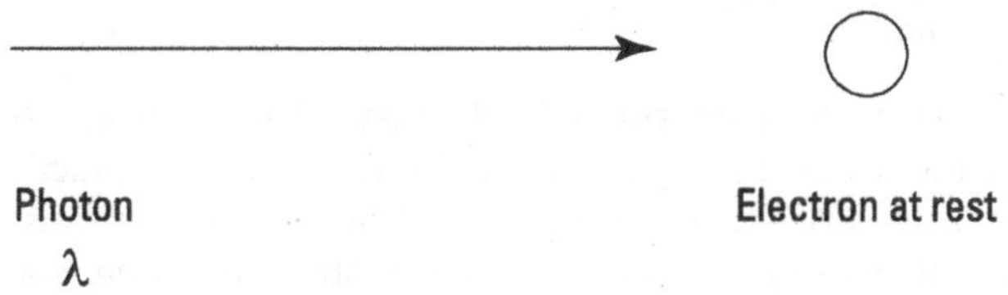

FIGURE 3-5: Light incident on an electron at rest.

Photon
λ

Electron at rest

Incident light (any light that falls on an object) comes in with a wavelength of λ and hits the electron at rest. After that happens, the light is scattered, as depicted in Figure 3-6.

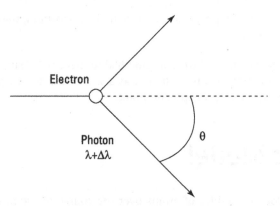

Electron

Photon
$\lambda+\Delta\lambda$

θ

FIGURE 3-6: Photon scattering off an electron.

Classically, here's what should've happened: The electron should've absorbed the incident light, oscillated, and emitted it — with the same wavelength but with an intensity depending on the intensity of the incident light. But that's not what happened — in fact, the wavelength of the light was actually changed by $\Delta\lambda$, called the wavelength shift. The scattered light had a wavelength of $\lambda + \Delta\lambda$ — in other words, its wavelength had increased, which means the light had lost energy. And $\Delta\lambda$ depends on the scattering angle, θ, not on the intensity of the incident light.

REMEMBER

Compton could explain the results of his experiment only by making the assumption that he was actually dealing with two particles — a photon and an electron. That is, he treated light as a discrete particle, not a wave. And he assumed that the photon and the electron collided elastically — that is, that both total energy and momentum were conserved.

Assuming that both the light and the electron were particles, Compton then derived this formula for the wavelength shift (it's an easy calculation if you assume that the light is represented by a photon with energy $E = h\upsilon$ and that its momentum is $p = \dfrac{E}{c}$):

$$\Delta\lambda = \frac{h}{m_e c}(1 - \cos\theta)$$

where h is Planck's constant, m_e is the mass of an electron, c is the speed of light, and θ is the scattering angle of the light. If you want to practice your trigonometry, you may notice that the structure of this equation means it can be rewritten in some other forms, such as with sine instead of cosine. You'll also sometimes see this written in a slightly different form, with the Compton wavelength of the electron, λ_c, defined as $\lambda_c = h / m_e c$.

In addition to earning the 1927 Nobel Prize in Physics for this work, Compton also began using the term *photon* to describe the light quanta in 1926. Although he wasn't the first to have done so, his adoption of this terminology led to it becoming widely used among physicists, and eventually becoming the official name for these light particles.

To derive the wavelength shift, Compton had to make the assumption that in this case, light was acting as a particle, not as a wave. And so, the particle nature of light was the aspect that was predominant.

Bohr's Atomic Model

Physicists adopted the idea of atoms over the course of the nineteenth century, but they still didn't have a clear picture of how these minute objects were constructed. The electron had been discovered in 1897, and scientists reasoned that the atom contained a positive charge. But the idea of a nucleus containing positively charged particles wasn't widely adopted yet.

The physicist Ernest Rutherford (1871–1937) proposed that the electrons moved around the center of the atom in orbits — much like the planets move around the Sun. If you've seen a traditional image of an atom with electrons moving in these sorts of orbits, you're looking at the Rutherford model of the atom.

But obvious problems with this model existed — most notably that since the electron was negatively charged and the nucleus was positively charged, physicists couldn't figure out how the electrons would stay in stable orbits. Newton had figured out how planets stay in orbits around the Sun (because of gravity), but someone needed to work out some similar explanation for the orbits of electrons around the atomic nucleus.

The changing of electron orbits

Enter young Niels Bohr, fresh off his 1911 doctoral dissertation showing that classical physics was falling short on explaining the electromagnetic properties of metals. He tackled this new challenge by applying quantum theory as a way to explain the electron's orbital behavior.

Bohr would later claim that the idea to apply the quantum wasn't a unique insight of his own. Physicists throughout Europe were, after all, talking about the new quantum theory, and many prominent physicists advocated that the theory would be useful to explain microscopic properties. And so, the thought that quantum theory might help resolve this electron-orbit problem was only natural.

The insight that inspired Bohr came from spectral lines (which I mention in Chapter 2). Spectral lines represent the energy emitted when an electron within an atom changes its orbit to a new orbit. In fact, an 1885 equation by Johann Balmer identified a pattern in the wavelengths of these lines by relating them to integer values.

Bohr realized that a continuous value like a wavelength being related to integer values sounded an awful lot like quantized behavior. He was able to use Balmer's equation to calculate what was going on with the electrons in the atom, including these activities:

>> **Absorbing and releasing photons:** When an electron in an atom is struck by a photon, it absorbs the energy of the photon by moving up to a higher orbit. When it later collapses down to a lower orbit, the electron will release a photon (which is what causes the spectral lines).

>> **Obeying the orbital rules:** But because the energy is quantized (only certain energies are allowed) and energy is related to the orbit, only certain orbits are allowed. The electron must move from one orbit to another instantaneously, which signifies another discontinuity introduced by the quantum theory.

TIP

The changing of electron orbits isn't the same as moving a satellite from one orbit around the Earth to another orbit. The satellite's transition isn't instantaneous. You actually move the satellite through intervening space. That is distinctly *not* allowed by Bohr's model, which would work only if the satellite teleports, jumps, or leaps from one orbit instantly into another distant orbit, without ever crossing the vast intervening space. *Note:* Vast is relative to the tiny size of the electron (or, metaphorically, the satellite).

Explaining results by using Bohr's model

REMEMBER

As weird as Bohr's idea was, it had the benefit of actually explaining the behavior observed in the simple hydrogen atom. Applying this concept to more complex atomic structures required more work, but in time, Bohr and his colleagues showed that this approach worked to explain the experimental results. And in science, fitting the experimental results can take you a long way toward becoming accepted — no matter how weird the idea is.

A Dual Identity: Looking at Particles as Waves

In 1923, the physicist Louis de Broglie suggested that not only did light waves exhibit particle-like aspects, but also that the reverse was true — all material particles should display wavelike properties.

How does this dual identity work? For a photon, momentum $p = h\upsilon / c = h / \lambda$, where υ is the photon's frequency and λ is its wavelength. And de Broglie defined the vector representing the wave, called the *wave vector*, as $k = p / \hbar$, where $\hbar = h / 2\pi$. De Broglie said that the same relation should hold for all material particles. That is,

$$\lambda = \frac{h}{p}$$

$$k = \frac{p}{\hbar}$$

De Broglie presented these apparently surprising suggestions in his PhD thesis. To test them, researchers turned to a classic experiment done with light: the double slit experiment. This experiment had actually provided some of the strongest evidence that light behaved as a wave (as discussed in Chapter 2). In Figure 3-7, you can see the setup and the results for this new quantum variant, where you use a beam of electrons instead of a beam of light. In this case, the screen is replaced by a film that changes color when an electron strikes it, allowing you to measure the intensity of electrons striking any portion of the screen.

If you have a single slit, the electron beam does exactly what you'd expect: It hits the screen on the other side of the slit. This can be seen in Figure 3-7 (a and b).

It's only when there are two distinct slits (as in Figure 3-7c) that things get interesting. You might expect the same result that you see with each individual slit, added together, but that's not the result that you get. Instead, the electron beam creates an interference pattern of intense and weak bands, which is a sign that the two light waves are overlapping. Thus, the experiment strongly confirms that this electron beam is exhibiting wavelike behavior, and not acting at all the way you'd expect particles to behave. These particles showed the exact same wavelike behavior that was seen with light!

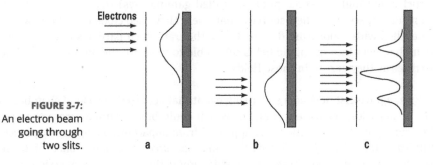

FIGURE 3-7:
An electron beam going through two slits.

The result was a validation of de Broglie's concept of matter waves. Experiment bore out the relation that $\lambda = h / p$, and de Broglie was proven a success. Award that man a doctorate!

REMEMBER

This dual identity of matter is probably one of the most counterintuitive and difficult concepts to grasp in all of quantum physics. In classical physics, particles are particles and waves are waves, and the distinction between them is clear. The findings of Einstein, Compton, Dirac, and de Broglie described in this chapter shattered that simple distinction in the era of quantum physics. Whether an object is behaving like an individual particle or a wave depends largely on how the scientist chooses to look at it. This feature of quantum physics is called *wave-particle duality*. Throughout the book, I commonly talk about matter in terms of waves, though — so get used to it.

Proof Positron? Dirac and Pair Production

In 1928, the physicist Paul Dirac posited the existence of a positively charged anti-electron, the *positron*. He did this by taking the newly evolving field of quantum physics to new territory by combining relativity with *quantum mechanics* (which deals with the mathematical description of the motion and interaction of subatomic particles) to create *relativistic quantum mechanics*. When he did this, though, his resulting equation gave an unexpected — and weird — result: a particle that seemed exactly like an electron but had a positive electrical charge.

The equation made a bold prediction: an *antiparticle* of the electron! And just four years later, physicists actually detected the positron.

In those days before particle accelerators, physicists relied on cosmic rays — those particles and high-powered photons (called gamma rays) that strike the Earth from outer space — as their source of particles. They used cloud chambers, which were filled with vapor from dry ice, to see the trails such particles left. They put their chambers into magnetic fields to be able to measure the momentum of the particles as they curved in those fields.

In 1932, a physicist noticed a surprising event that occurred in a cloud chamber. A pair of particles, oppositely charged (which could be determined from the way they curved in the magnetic field) appeared from apparently nowhere. *No particle trail led to the origin of the two particles that appeared.* The appearance of these particles from nowhere was called pair production — the conversion of a high-powered photon into an electron and positron, which can happen when the photon passes near a heavy atomic nucleus.

So, experimentally, physicists had now seen a photon turning into a pair of particles, which provided further evidence of the particle nature of light. Later on, researchers also saw pair annihilation: the conversion of an electron and positron into pure light.

Pair production and annihilation turned out to be governed by Einstein's newly introduced theory of relativity — in particular, his most famous formula, $E = mc^2$, which gives the pure energy equivalent of mass.

You Can't Know Everything (But You Can Figure the Odds)

Probability is always a measure of uncertainty, as discussed in Chapter 2. In quantum physics, as various experiments show, it isn't even certain whether a wave and particle are different things! Quantum mechanics lives with an uncertain picture quite happily. That view offended many eminent physicists of the time — notably Albert Einstein, who said, famously, "God does not play dice."

This opinion is particularly ironic because in 1916, Einstein wrote one of the first papers to apply probability in a quantum context. In this paper, he used probability as a way of interpreting the atomic transitions from the Bohr model of the hydrogen atom. (See the section, "Bohr's Atomic Model," earlier in the chapter.) But years later, Einstein objected to the full scope of how probability manifested in the quantum realm.

In this section, I discuss the idea of uncertainty and how quantum physicists work in probabilities.

Position versus momentum: The Heisenberg uncertainty principle

The fact that matter exhibits wavelike properties gives rise to more trouble — waves aren't localized in space, and so boundaries that are treated as clear and well-defined in classical physics cease to be either in the quantum realm. And knowing that inspired Werner Heisenberg, in 1927, to come up with his celebrated uncertainty principle.

You can completely describe objects in classical physics by their momentum and position, both of which you can measure exactly (within the precision range of your measuring instrument, that is). In other words, classical physics is completely deterministic.

Understanding uncertainty when taking measurements

On the atomic level, however, quantum physics paints a different picture. Heisenberg realized that the more precisely you tried to measure an object's position, the less precisely you could measure the object's momentum (or speed). And vice versa.

Consider this analogy for measuring position versus momentum: If a race car were zipping down the street, you could focus your attention on measuring either the moment it hit a precise point on the track (position) or the time it took for it to go a certain distance (its momentum or speed). The more focus you're trying to place on one of those measurements, though, the less attention you're going to be able to give to making the other one.

But the uncertainty in quantum physics doesn't just occur because scientists don't have the focus or tools necessary to make a precise measurement. This point really cannot be emphasized enough. The reason for the uncertainty in quantum physics stems from wave-particle duality. You cannot precisely measure a wave the way you can a feature of a speeding car. You can't pin down a wave, no matter how good your measurement instrument is.

Quantifying uncertainty

Here, the *Heisenberg uncertainty principle* says that an inherent uncertainty exists in the relationship between position and momentum, and not in a vague way. The uncertainty precisely quantifies the relationship. Heisenberg did the math to back up his claim, which was subsequently borne out by experiment. (I walk through the steps to derive the Heisenberg uncertainty principle in Chapter 7.) If Δx is the measurement uncertainty in the particle's x position, then Δp_x is its measurement uncertainty in its momentum in the x direction, and he defined a new constant, $\hbar = h / 2\pi$ (usually spoken of as h-bar or the reduced Planck's constant), to get this formula:

$$\Delta x \Delta p_x \geq \frac{\hbar}{2}$$

Note that the two uncertainties (Δx and Δp_x) are multiplied, and they must be greater than a constant number (albeit, to be fair, a very small constant number). If you were making measurements of these two traits that were as precise as physically possible, and you wanted to cut one of the uncertainties in half, then you could only do it by doubling the uncertainty in the other value. Something's gotta give.

Quantum physics, unlike classical physics, is completely indeterministic. You can never know the precise position and momentum of a particle at any one time. You can give only probabilities for these linked measurements.

Applying uncertainty to particle physics

Another variant of Heisenberg's uncertainty principle applies in the case of particle physics and is specifically useful in talking about particles that decay within a certain lifetime. That equation takes the form

$$\Delta E \Delta t \geq \frac{\hbar}{2}$$

where ΔE is the uncertainty in the measurement of the energy, and Δt represents the uncertainty in the lifetime of the particle, and $\hbar$ is the reduced Planck's constant described earlier in this section.

This equation comes up in a number of cases in quantum physics. Consider a common example: When atoms drop out of excited states, they emit light that forms emission spectra. Observing the atomic emission spectra for these atoms reveals a series of spectral lines, which relate to the energy states of the transition. The line is always a spread of emissions across a range, representing the variance in the energy of the state change.

What the time-energy version of the uncertainty principle says is that multiplying the uncertainty in the energy times the uncertainty in the time must be greater than half of the reduced Planck's constant. If you have a very small possible Δt variation in time (the excitation is very short-lived), then you'll have a larger ΔE (broader spectral lines). And, following the same logic, excited states that last for a longer period of time (larger Δt) will have narrower spectral lines (smaller ΔE).

Rolling quantum dice: A new take on probability

In quantum physics, the state of a particle is described by a wave function, $\psi(r, t)$. The wave function describes the de Broglie wave of a particle, giving its amplitude as a function of position and time. (See the earlier section, "A Dual Identity: Looking at Particles as Waves," for more on de Broglie.)

REMEMBER

The wave function gives a particle's amplitude, not intensity; if you want to find the intensity of the wave function, you have to square it: $|\psi(r, t)|^2$. The intensity of a wave is the characteristic that's equal to the probability that the particle will be at that position at that time.

Two approaches to outcome probabilities

This insight into the proper interpretation of the wave function was recognized by Max Born, who was actually Heisenberg's mentor. Born recognized in 1925 that quantum physics required a radical new approach to mechanics, and he set his protégé Heisenberg on the path that led to the creation of the original iteration of quantum mechanics. At crucial points, he helped guide Heisenberg through confusing calculations to reach the correct form of quantum mechanics.

But in both Heisenberg's matrix form of quantum theory and the 1926 wave function approach developed by Schrödinger, Born recognized a common theme: The resulting outcomes represented probabilities, not absolute, deterministic predictions like those expected from classical physics. And Born further recognized that the way that you get that probability is by squaring the wave function.

Exploiting a wave function

That's how quantum physics converts issues of momentum and position into probabilities: by using a wave function, whose square tells you the probability density that a particle will occupy a particular position or have a particular momentum. In other words, $|\psi(r, t)|^2 d^3r$ is the probability that the particle will be found in the volume element d^3r, located at position r, at time t. (*Note:* The d^3r are not variables here, but calculus notations.)

Besides the position-space wave function $\psi(r, t)$, you can also have a momentum-space version of the wave function: $\phi(p, t)$.

REMEMBER

Leaving the mathematics notation aside, this means that the equation (the wave function) is directly related to how probable it is that a physicist will see any specific outcome when they look for that outcome. That is the mathematical result you get out of quantum physics: a wave function that gives you a probability. In classical physics, probability is a sign of uncertainty, but in quantum physics, probability is the thing that you can be most certain about. To get all of this to give you concrete answers, you need to use the symbols and perform calculus to integrate the probability over the region of interest. (I get to the particulars of the notation in Chapter 7, for those who want to dive into the math.)

This book is largely a study of the wave functions of various types and states of particles, including the wave functions of

>> Free particles

>> Particles trapped inside potentials

>> Identical particles hitting each other

>> Particles in harmonic oscillation

>> Light scattering from particles

Using quantum physics, you can predict the behavior of all kinds of physical systems . . . which I show you in Part 3 of this book.

**TECHNICAL
STUFF**

INTERFERENCE AND SUPERPOSITION REVISITED

The behavior of particles and other objects in quantum physics is defined by the wave function. While you find more details about the wave function throughout this book, one outcome of it — interference — shows up in my discussion of the double slit experiment earlier in this chapter. (See the earlier section, "A Dual Identity: Looking at Particles as Waves," and Chapter 2 for a discussion of wave interference.)

The same thing is happening with the wave function of matter in the double slit experiment: For two matter waves, $\psi_1(r, t)$ and $\psi_2(r, t)$, the phase difference between them is what actually creates the interference.

Superposition of waves is the process of uniting the two waves into one single combined wave — or, in quantum physics, a wave function. If you have two wave functions, you can add their amplitudes, not their intensities, to sum them into a single superposed wave function:

$$\psi(r, t) = \psi_1(r, t) + \psi_2(r, t)$$

Quantum physics gets weird, compared to classical physics, precisely because concepts that are well defined in classical physics, like position and momentum, become a strange superposition of possible quantum states.

A New Take on Light: Quantum Electrodynamics

One of the outcomes of quantum physics experimentation was the realization that light moved in small, discrete, quantized packets of energy, or photons. But throughout the nineteenth century, physicists had clearly adopted the idea of light as a wave, manifesting most explicitly in the electromagnetic waves described in Maxwell's equations. How could both of these interpretations be possible?

The quest to resolve this inconsistency ultimately resulted in the development of *quantum electrodynamics*, or QED, which is the shorter name for the relativistic quantum theory of electrodynamics. This theory is the quantum counterpart to Maxwell's classical electromagnetic theory. Quantum electrodynamics explains the interaction between matter and light, and electrically charged particles.

What is doing the waving?

In the late 1800s, experimental physicists tried to confirm the wave interpretation of light and electromagnetic waves by detecting the *ether* (the medium that transported the waves). Since they couldn't detect it directly, they attempted to detect it indirectly, by identifying the expected influence of the ether's movement on the path of light. The most notable researchers in this area were Michelson and Morely, whose experiments I discuss at greater length in Chapter 2. The short story: Their efforts were unsuccessful because the ether that they were looking for didn't actually exist!

REMEMBER

While quantum physics eliminated the need for the ether, it also offered up a new explanation for the wavelike behavior. One of the earliest innovations in quantum physics was the understanding of the duality of waves and particles. Not only did light take on properties of both waves and particles, but so did matter. (See the section, "A Dual Identity: Looking at Particles as Waves," earlier in this chapter for the wave interpretation of matter.)

But what is actually doing the *waving* in an electromagnetic wave? Well, one answer is inherent in the foundation of quantum theory: the quantum wave function itself. Since the wave function represents the probability that a photon will be in a given position, when you're looking at the double slit experiment (see Chapter 2), one interpretation involves thinking of the wave function as a wave that moves through both slits. This wave is then calculated as a superposition of two waves, each going through a different slit, and the location of the photons becomes an interference pattern of those two wave functions. This interpretation means the probability of the photons arriving at the screen would rise and fall in a way that matches the interference pattern of waves.

Trying to understand the wave interpretation of light is one of the many non-intuitive outcomes of quantum physics, so if you aren't quite following the description given here, then you're in good company. Einstein, among others, also thought this idea sounded like nonsense, even while acknowledging the overall experimental success of quantum theory. The meaning of this interpretation, and disputes about it, are the focus of Chapter 6.

First glimmers of QED

Ultimately, a disconnect between Maxwell's equations and quantum physics theories emerged. Maxwell's equations described the electromagnetic wave as a continuous form of energy, and in quantum physics *nothing was continuous*. Energy was always quantized and discrete when viewed at quantum levels. While the energy might manifest in accordance with Maxwell's equations when viewed at a larger scale, the fundamental physical description of what happens had to fit the quantum picture. In the aftermath of World War II, physicists turned their attention toward reconciling the differences between the two approaches (continuous versus discrete).

One of the first physicists to tackle this challenge was Paul Dirac, who I mention in the section, "Proof Positron? Dirac and pair production," earlier in this chapter. He was one of the earliest physicists to try to reconcile quantum physics with Einstein's theory of relativity. In a 1927 paper, Dirac coined the term *quantum electrodynamics* (QED) to describe this relationship, shortly after quantum mechanics was originally developed, as discussed throughout this chapter.

Here's the problem that Dirac recognized in 1927: The existing classical electromagnetic field theory from the Maxwell equations didn't match up with quantum theory. Due to the uncertainty principle, quantum harmonic oscillators (see Chapter 9) cannot be completely stationary. They must always be moving at the quantum level and maintaining a minimum amount of energy.

REMEMBER

Just as physicists had to update classical mechanics into quantum mechanics to explain the behavior of objects moving in the quantum realm, they also needed to update classical electromagnetic theory into a new type of quantum field theory to explain quantum behavior in the electromagnetic field.

The problem that quantum physics encountered when trying to transition electromagnetic theory was in getting precise answers. They might get very good approximations of answers by using a method called *perturbation theory*, but ultimately, trying to make their calculations more precise eventually led to divergent terms that went to infinity. Because nothing in the physical world actually has infinite energy, these infinite terms couldn't possibly represent physical results. Electromagnetic theory at a quantum level needed a way to legitimately resolve the results.

PHYSICS FIELDS REVISITED

In a Chapter 2 sidebar, I introduce the idea that classical physics evolved from Newton's perspective of looking at individual point masses acted on by forces and, instead, transitioned to looking at more holistic fields. Basically, the same transition happened in quantum physics when physicists reinterpreted the original quantum mechanics formulations into a more comprehensive *quantum field theory* (QFT).

As mentioned in this chapter's section, "First glimmers of QED," Paul Dirac noticed the need for this transition in 1927, when he recognized the need for *quantum electrodynamics*, a quantum theory for the electromagnetic field. Ultimately, quantum field theory became more general because physicists recognized the existence of other types of quantum fields, such as those of the strong and weak nuclear force.

You may notice a real conflict between the way of thinking that goes into these two approaches. Feynman (see the next section, "The photon gets a new job"), for example, understood the field approach, but seemed to be much happier speaking in terms of particles. Schwinger was always frustrated by Feynman's visual approach for precisely this reason; he felt that it prompted people to think about the particles rather than the quantum field.

The gravitational field is the one field that physicists have never successfully reconciled as a quantum field theory. And the search for a gravitational field has been the holy grail of physics for a century. Einstein devoted the last half of his life to this unsuccessful search. For a much deeper look at the search to reconcile quantum physics and gravity, I humbly recommend *String Theory For Dummies* (Wiley, 2022).

The photon gets a new job

One physicist who thought long and hard about this problem of quantum physics and waves was Richard Phillips Feynman. You can find entire books devoted to Feynman (several of them by Feynman himself), and while all of the quantum physicists in this book had interesting lives, I see something particularly energizing about Feynman's. If you want to learn more about him, I direct you toward his entry in Chapter 16. For now, I focus specifically on his role in developing quantum electrodynamics.

Visualizing with Feynman's diagrams

Feynman spent a decent chunk of the 1940s helping to develop the nuclear bomb as a scientist on the Manhattan Project. (At the same time, he taught himself to be an adept safe cracker. Like I said, he's an interesting character.) Before that, though, he had time to complete his doctorate in 1942, and in that doctorate he laid out some core ideas that would be the basis of quantum electrodynamics.

One idea in his doctoral thesis became known as *Feynman diagrams*, a visual approach to representing quantum interactions. In these diagrams, Feynman represented particles using lines of various types. Particles such as electrons moving through space show up as straight lines. Photons are wavy lines. The axes of the graph represent space in the horizontal direction and time in the vertical direction, so they're clearly an abstraction and don't precisely define all of spacetime; however, the diagrams are tools for thinking about the interactions of particles. Examples of such an interaction appear in Figure 3-8.

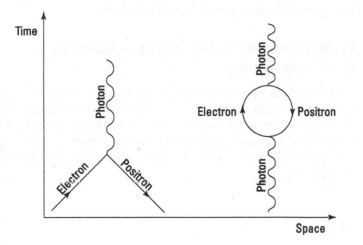

FIGURE 3-8:
(Left) A particle and antiparticle annihilate each other, releasing a photon. (Right) A photon splits into a particle and antiparticle, which immediately annihilate each other.

The Feynman diagrams in Figure 3-8 are related to antimatter, the electron, and its antiparticle, the positron, which I discuss earlier in this chapter, in the section, "Proof Positron? Dirac and pair production." So just what do these images communicate? Feynman's insights, which he depicts both visually and mathematically, include

>> **The positron can be represented as if it were an electron moving backward in time.** The electrons are depicted as arrows moving forward in the time direction, while the arrows on positrons are similar except that the arrows move in the opposite direction.

>> **Colliding protons and electrons create photons.** The image on the left (Figure 3-8) shows an electron and a positron moving through space. When the electron and positron collide, they then cease to exist, and a photon is formed from the energy of their annihilation of each other. That photon then moves forward in time. Pretty cool, huh?

>> **Virtual particles can exist for brief periods of time.** The image on the right shows that, in the photon's place, both an electron and a positron have formed. These *virtual particles* exist for a mere moment, but are drawn together, since they are opposite charges. They collide with each other, and then annihilate each other. Now, again, a new photon is moving forward from that point.

Earlier in this chapter (in the section, "Position versus momentum: The Heisenberg uncertainty principle"), I mention that uncertainty in time is related to uncertainty in energy by Heisenberg's uncertainty principle. One way to think about the creation of virtual particles is this: If the time period is short enough, virtual particles can "borrow" energy from the universe to exist. But that thinking only works if the time period is *incredibly* short.

Tying Feynman's diagrams to quantum electrodynamics

So how do these Feynman diagrams tie into the work developing quantum electrodynamics? Figure 3-9 is the best indication of the connection; it shows two electrons nearing each other. Not surprisingly, because they both have a negative charge, they push against each other and begin moving in the opposite direction. Fine, that's what you'd expect. However, what's that squiggly line in the middle? It's a photon.

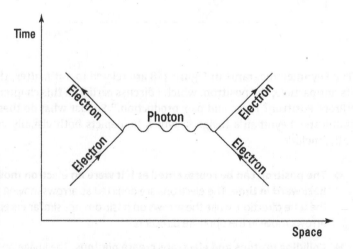

FIGURE 3-9:
A Feynman diagram demonstrates how two particles interact with each other.

REMEMBER

Classical electrodynamics says that the electromagnetic force between any two objects, such as two electrons, is carried in the form of an electromagnetic wave — and an electromagnetic wave is a type of light wave. (See Chapter 2 for more on this subject.)

So, if the two electrons are interacting due to an electromagnetic force between them, that force is going to be carried by a photon. In fact, what Feynman's diagram (and the more rigorous mathematical work in his doctoral thesis) predicted was that a photon would spring into existence for a brief instant for the express purpose of communicating between these two particles. It is a *virtual photon*.

TIP

In fact, in quantum electrodynamics, the electromagnetic force is said to be *mediated* by the photon. The photon is the particle that communicates the information about the electromagnetic field, and thus the electromagnetic force, to the various entities within the field. And that's exactly what is shown in Figure 3-9.

Feynman wasn't alone in his work in this area, of course. But most of the physicists were approaching it using field operators, a more traditional approach to the concept. Among the physicists who would make some of the more groundbreaking, related discoveries were Julian Schwinger, Sin-Itiro Tomonaga, Freeman John Dyson, and Murray Gell-Mann.

It was Dyson who showed in 1949 that these two approaches — the graphical method by Feynman and the more operator-driven method by the other researchers — were equivalent. Schwinger, Tomonaga, and Feynman would go on to share the 1965 Nobel Prize in Physics for this research.

Breaking Open the Atom's Bits

Another series of physics revolutions moves through the early portion of the twentieth century, reaches its culmination at the climax of World War II, and then continues onward from there. This series is not a parallel track to the development of quantum physics. These developments are more like two long paths moving through the same woods, winding around and crossing each other in multiple places, perhaps even joining together into a single path for a time.

The discovery of quantum theory had to do with atoms emitting energy, and Bohr's use of quantum theory to describe atomic structure really gave momentum to the adoption of both quantum theory and atomic theory. Quantum insights were leading to the discovery of new particles, including smaller structures within the atom, like the positively charged proton and the non-charged neutron.

This growing understanding of the atomic structure, made possible by quantum theory, meant that physicists could begin to explain properties such as radioactive decay. In Chapter 2, I mention that the creator of the periodic table of the elements, Dmitri Mendeleev, didn't even believe in atomic theory. Now the deep structural properties of the different atomic elements were clearly understood as explaining the organizational structure of that very table.

Banging particles together to get information out

Many of the physicists already discussed in this chapter gathered as part of the Manhattan Project to build the first atomic bomb (or, in the case of Heisenberg, to work on the German equivalent). And while that effort resulted in death and destruction, it also laid a path for exploration and knowledge, with the growth of particle accelerators in the wake of World War II.

In a particle accelerator, as the name suggests, physicists use powerful, focused magnetic fields to accelerate charged particles. And then, once they have them moving fast enough, they smash the particles into something. The *something* can be either other fast-moving particles, or maybe some stationary object. And they observe what happens.

I worked at a particle accelerator during a summer internship at Indiana University Bloomington. My work involved colliding a beam of highly charged particles into a piece of film that changed colors when exposed to radiation, and then using that color change to calculate the amount of radiation. The university also used the accelerator to test satellite parts by observing how the circuitry reacted when blasted with charged particles. (Pieces of calibrated film from my experiments, meanwhile, accompanied the satellite parts in the accelerator so that they could check the film pieces later to provide a more precise measurement of the radiation at different points in the device.)

REMEMBER

In the years after World War II, physicists learned a lot from banging high-speed particles and atoms together. They learned so much, in fact, that they discovered entirely new types of forces that they hadn't known about within the atom.

Holding the atomic nucleus together

Before getting too far into atomic forces, remember that what you already know about the atom is useful. For example, you know that atoms have a nucleus that is positively charged, and that the nucleus is circled by moving electrons. The Bohr model of the atom (see the section, "Bohr's Atomic Model," earlier in this chapter) explained why those electrons didn't collapse into the positively charged nucleus.

As soon as physicists began looking more closely at atomic structure, they realized that the atomic nucleus contained

>> **Protons:** Positively charged particles

>> **Neutrons:** Neutral (no-charge) particles

Discovering these subatomic particles and the cohesion of the nucleus made it clear that some sort of force within the atom itself held the nucleus together. The positively charged protons *should* be pushing against each other. And because they're so close together, the force of that pushing should be incredibly powerful. Something had to be holding these protons and neutrons together, with enough power to overcome the electromagnetic forces between the protons.

In fact, research would show that two nuclear forces exist within the atom.

>> **Weak nuclear force:** An interaction that binds particles together and is about 100,000 times weaker than the electromagnetic interaction.

   The weak nuclear force is related to particle decay and is mediated by particles that have the particularly uninteresting names of *W bosons* and *Z bosons*.

>> **Strong nuclear force:** An interaction binding protons and neutrons together, which is significantly stronger than any of the other known forces.

   Similar to the electromagnetic force, the strong nuclear force is also mediated by a particle. (See the previous section for more info.) But instead of a photon, the particle that communicates the strong nuclear force is a messenger particle called a *gluon*. These gluons don't just hold protons and neutrons together; they also work within the protons and neutrons, to hold together the protons' and neutrons' individual parts, which are called *quarks*!

REMEMBER

The strong nuclear force is really the one here that answers the question about what holds the atomic nucleus together. When two protons are close together to each other, or to a neutron, they get bound together by this intense force, which overwhelms the repulsive force from electromagnetism.

With quantum field theory established, the work in quantum physics through the second half of the twentieth century focused on experimental work to figure out all of the bits and pieces. This work resulted in the Standard Model of Particle Physics, which was considered completed in 2013, with the discovery of a particle called the *Higgs boson*. I provide more details on all of that in Chapter 5.

2

The Fundamentals: Quantum Physics Principles and Theories

Find out what quantum physics says about particle states.

Look back at how quantum physics has revolutionized our understanding of electromagnetism and matter.

Investigate the central interpretations, contradictions, and debates of quantum physics.

Chapter **4**

Quantum Mechanics: Particle States and Dualities

I n the field of quantum physics, you tend to look at individual particles and how those particles behave. (You dive a little bit into multiple particle systems in Chapter 15.) At the core of your examination is understanding that quantum physics defines individual particles in terms of their particle states. Suppose that you have a particle that is in a certain state at one point; examining the particle further means you look at the likelihood that the particle will be in certain other states at other points.

In this chapter, you find out about quantum states of particles. I start with an explanation of the quantum numbers that physicists use to describe quantum states, and some specific ways those states manifest in physics. Then I return to an idea, introduced in Chapter 3, of wave-particle duality, which is central to quantum physics. Finally, I talk a bit about how quantum physics led physicists to predict and then discover antimatter.

Quantifying by Quantum Numbers

Particles in quantum physics are in defined states, and these states are represented by a sequence of numbers. The numbers that quantum physicists use to define a quantum state are called *quantum numbers*.

Niels Bohr first used this approach when he developed the Bohr model of the atom, as described in Chapter 3. He was trying to describe the physical state of electrons within a hydrogen atom. His initial approach included just a single quantum number, n, which started at 1, and then increased by integer values, so you would say $n = 1, 2, 3, \ldots$, to describe the state of the electrons.

Bohr's model, although it explained quite a lot, wasn't complete. Research in subsequent years showed that the structure of electrons within an atom was quantum mechanically even more nuanced than Bohr had originally thought. These nuances took two additional quantum numbers — ℓ and m — to describe the electrons' structure.

You see the quantum number n at work in the equations that show up starting in Chapter 8, but quantum numbers ℓ and m don't show up again until Chapter 10.

The ever-spinning electron

Another quantum number is at work in all particles. This number isn't dependent upon the physical position of the particle within the atom but, instead, is a more intrinsic property — called *spin* — and it's denoted by the quantum number s.

In short, *spin* represents a sort of intrinsic angular momentum within the particle. I devote all of Chapter 11 to spin, so go there if you want to find out how it was discovered, along with more details on how to work with it. But I devote a little time in this chapter to explaining the concept, because it is central to particle physics.

Imagine that you have a spinning basketball, say, on the finger of a basketball player. This ball has an angular momentum. The basketball player, by spinning the ball a certain way, can increase or decrease the spin of the ball, and thus increase or decrease the ball's angular momentum. That's a classical way of looking at the concept of spinning something.

Quantum physics tells you that an electron — or any fundamental particle — has a different relationship with spin. Every electron is identical to every other electron, and they're essentially point particles. In other words, there's really no way to "spin" an electron like you can a basketball. And so, you can't increase or decrease the angular momentum in this way. The only angular momentum an electron has is related to the spin that's intrinsic to the atom.

TIP

Moreover, the electron's spin is always a constant value, designated as ½. But the direction of the spin can be in one of two ways, either up spin or down spin, represented as $s = +\frac{1}{2}$ or $s = -\frac{1}{2}$. It's like you have a basketball that spins at exactly one rotation per second, either clockwise or counterclockwise. There's no slowing down, stopping, and going the other way; it's either one way at that speed or the other way. This completely crazy idea is what quantum physics tells you. (The details about the quantum physics of angular momentum are in Chapter 10 — and spin is in Chapter 11 — in case you want to dig deeper into the experiments and math that led to this crazy conclusion.)

Fermions and bosons

The reason that makes spin such an essential property to come out of quantum physics involves the number of particles that particle physicists have discovered. When the atom was explored in particle accelerators, the fact that its interior structure was much more complex than previously believed became apparent.

REMEMBER

While electrons within an atom are standalone particles that cannot be broken down any further, the protons and neutrons that make up the atomic nucleus can, in fact, be broken down into even smaller particles. And these smaller particles from the protons and neutrons weren't the only particles discovered. By using particle accelerators, physicists found a wealth of new particles, some of which drift around in outer space, and others that exist only for mere fractions of a nanosecond in a laboratory before disintegrating.

What became evident, though, is that these particles have a structure that is heavily tied to their spin. The particles that have a half-integer spin, like the electron, behave differently (in key ways) from those particles that have an integer spin, like the photon.

The half-integer spin particles are called *fermions*. The particles that have spin of a full integer are called *bosons*. When you have multiple particles, as I discuss in Chapter 15, these different types of particles behave very differently. I also offer more about the difference of these types of particles in Chapter 5.

Revisiting Wave-Particle Duality

One of the most unexpected discoveries of quantum physics was wave-particle duality. First, physicists discovered that light behaves not just as a wave, but also as a particle. Then they discovered that particles, like electrons, also behave like waves! These unexpected behaviors don't show up all the time, but they do so when you carefully construct an experiment that can show the appropriate behavior — for example, like the double slit experiment I discuss in Chapters 2 and 3.

Origins of wave-particle duality

It was Louis de Broglie who found that matter particles could also behave like waves, creating the equation that defined the *de Broglie wavelength*, λ, for the matter wave:

$$\lambda = \frac{h}{p}$$

In this equation, the h is the Planck constant, but the p is the momentum of the particle. The momentum, p, of the particle is the mass, m, times the velocity, v, and $p = mv$, so you can also write this equation as

$$\lambda = \frac{h}{mv}$$

This equation actually shows why wave-particle duality is such a hard relationship to see. Consider these points as they relate to the mass, m, which is in the equation's denominator:

>> **When you look at particles with larger and larger masses** — larger values of m — the wavelength gets smaller until it is so small that you can never witness any wavelike properties of the matter.

>> **When you look at particles with smaller and smaller masses** — smaller values of m — the wavelength gets larger and therefore easier to see. Even then, you must carefully construct any experiments.

And, in fact, de Broglie introduced some ideas for such experiments into his 1924 doctoral thesis, where he proposed this equation and the concept of matter waves. His ideas were experimentally confirmed in 1927.

Implications of wave-particle duality

Wave-particle duality is conceptually important in quantum physics because the equation at the heart of quantum physics is called the *quantum wavefunction*, usually denoted as ψ (although sometimes other symbols are used). Even when you talk about a single particle, in quantum physical terms, you actually refer to the wave function that describes that single particle. Parts 3 and 4 (in this book) involve a much deeper exploration of the mathematics of working with this wave function.

Individual particle and wave aspects blend together in wave functions. When properly set up and analyzed, the wave function can be used to calculate the *probability* that a particle will be in a certain state or location at a given time. The calculation doesn't give certainty, but only a probability. That's why uncertainty is such a central feature of quantum physics, as described in Chapters 1 and 3.

But once you make the measurement to find where the particle is, you're no longer talking about the wave function. Now you're talking about an actual particle, which has an actual defined state. You know what particle state exists at the instant of measurement, and no longer must talk about it in terms of mere probability.

In Chapter 6, you find out how different interpretations of quantum physics approach this transition from a probabilistic wave function into a concrete value.

Discovering What Antimatter Is

Shortly after Werner Heisenberg came up with the original equations for quantum mechanics, a curious thing happened. The physicist Paul Dirac began working on a version of quantum mechanics that would include Einstein's theory of relativity, and came up with an equation that represented the electron. The problem was that the formula allowed the electron to have either a positive or negative charge, even though the electron *always* had a negative charge.

If this had been the end of the story, then Dirac would have just needed to modify his equation so that it only produced negative results. But a few years later, physicists actually detected the positively charged particle predicted by the equation! (See Chapter 3 for more on this.)

This newly discovered particle was called the *positron*, and it was exactly like the electron except that it had a positive electrical charge. The positron was an antiparticle of the electron, and the first example of antimatter.

REMEMBER

The growing research in particle accelerators showed that the positron wasn't the only antimatter particle in physics. One key feature of antimatter is that when it encounters ordinary matter, the matter and antimatter particles annihilate each other with a release of energy in the form of photons. (Since photons don't have an electric charge, there is no antiphoton . . . or, to think of it another way, the antiphoton is also just a photon. It is its own antiparticle.)

Chapter **5**

Quantum Electrodynamics and Beyond

After the basic ideas of quantum physics were laid out in the 1920s, World War II caused a bit of a distraction in research related to the field. The Manhattan Project team achieved a great (and horrible) success in the form of the atomic bomb, and the greatest physicists of a new generation — for whom quantum physics was well established before they even got to college — sought to explore the quantum realm even more deeply. They dedicated themselves to turning quantum physics into a theory that undergirded their understanding of virtually every aspect of the physical world.

In this chapter, you see how physicists throughout the middle of the twentieth century explained the structure of matter and energy (almost entirely) in terms of quantum physics. You find out about some of the inventions that came out of the quantum revolution — inventions that are so ubiquitous now that most people don't even realize how much they rely on them. Then you take a look at something that feels like science fiction: the amazing possibility of quantum computing.

Quantum Field Theory: Explaining Matter and Energy

Quantum physics studies brought about amazing achievements throughout the middle of the twentieth century and explained (at the most fundamental level) nearly everything about matter and energy. In Chapter 2, I describe classical physics ideas that took centuries to understand and refine. In the span of a few decades — starting when Max Planck first proposed quantizing physical properties in 1900 (see Chapter 3 for more about this proposal) — physicists used Planck's idea to explain the nature of electromagnetism, as well as the forces that held atoms together!

The foundation of these revolutionary ideas was *quantum field theory*, an approach to quantum physics that uses the concept of fields. A *field* in physics is a model where space is filled with a physical quantity that has a value at each point in space and time. If the field also has a direction, it's called a *vector field*. (Check out Chapters 2 and 3 for more about fields.)

TIP

Quantum field theory defines quantum physics in terms of a quantum field that permeates all space and has *energy* (a scalar or magnitude) and *momentum* (a vector with both magnitude and direction) throughout space and time. By using this definition, you can then view the quantum interactions as being related to the behavior of that field at a given point in space.

I don't take you too deeply into this theoretical approach, and even in the mathematics of Parts 3 and 4, I avoid diving directly into the mathematics of quantum field theory. But for now, I want you to see how this important concept shows up in quantum theory.

Revisiting quantum electrodynamics

Quantum electrodynamics, or QED, is the quantum theory that describes electromagnetism. I cover the history of its development in Chapter 3, but here I dive a bit more deeply into how electromagnetism actually works.

One insight from Maxwell's equations (see Chapter 2) is that electricity and magnetism are related, unified forces, which can present as an electromagnetic wave that influences charged particles. In addition, the equations describe light as a form of electromagnetic wave.

The first insight relevant to electrodynamics from quantum physics is that light is made up of photons, so the natural consequence of that fact is that the electromagnetic wave governing electricity and magnetism is also made up of photons. In other words, photons do the work of generating electricity and magnetism, or electromagnetic force!

That realization is the starting point in understanding quantum electrodynamics. Consider one of the simplest cases of two electrons interacting with each other. They both have a negative charge, so classical physics tells you that they will repel each other. But why? Because the electromagnetic wave communicates the charge between them. In quantum physics terms, what happens is that the electrons generate and exchange photons. In terms of quantum physics, the photon *mediates* the electromagnetic force and serves as the particle that communicates information about the force between the electrons.

This key insight is attributed mainly to Richard Feynman and is part of what led to his Nobel Prize in Physics. Chapter 3 illustrates two particles interacting through a photon in a Feynman diagram, just as described in the preceding paragraph. And it wasn't just that he came up with cool graphics. Feynman also laid out the underlying mathematical basis for this description!

QED is immensely important because it describes electromagnetism — an incredibly important force — at the quantum level. Atoms are surrounded by shells of electrons in different quantum states, and when two objects come in contact, it's actually the outer shells of electrons that are the structures that interact with each other. Almost every physical interaction that takes place in our day-to-day lives is ultimately explained by electromagnetism!

Looking at nuclear forces and quantum chromodynamics

With the foundation of QED laid, physicists also began to explore the structure of the atom. They discovered that two additional forces beyond electromagnetism were at work within the atom. These forces are known as the weak nuclear force and the strong nuclear force (which I introduce in Chapter 3), and their names denote their strengths relative to each other.

You can also explain both forces in terms of quantum field theory. Instead of being mediated by photons, these forces are mediated by other types of particles, which are also bosons, like the photon. But the nuclear force particles are specifically called *gauge bosons* and exist as virtual particles that show up just for the purpose of communicating forces between particles in quantum field theory. (See the nearby sidebar "A Theory of Everything?" for a discussion of how this same idea may relate to the force of gravity.)

REMEMBER

The theory that describes the strong nuclear force, where particles called quarks interact through gauge bosons called *gluons*, is referred to as *quantum chromodynamics*, or QCD. This name comes from a quantum number, called *color*, that performs a role somewhat analogous to electrical charge.

The gluons perform two major functions in quantum chromodynamics:

» **Holding protons and neutrons together:** The protons within a nucleus exchange virtual photons (because they both have a positive charge), and those virtual photons tell the protons to electromagnetically repel each other (in accordance with QED). But they are also transferring virtual gluons, and those virtual gluons are telling them to bind together into an atomic nucleus.

 The strong nuclear force — mediated by these virtual gluons — is holding the protons (and neutrons) together with more power than the virtual photons are trying to repel them. The strong nuclear force is so much stronger than the electromagnetic force.

» **Holding quarks together to form protons and hadrons:** The proton is divisible. It actually consists of three different particles, called *quarks*, that are held together by the gluons!

TIP

You won't ever just stumble upon a stray gluon (such as the gauge boson for the strong nuclear force) just hanging out. It exists solely for the purposes of holding quarks together into protons or neutrons, or for holding the protons or neutrons together. These virtual particles have a job to do!

Calling on the Higgs boson

One last particle worth mentioning is the Higgs boson. It was predicted as a consequence of quantum field theory, as a particle needed to explain some asymmetries in the mathematical model. To use a simplistic description, the *Higgs boson* theory explains why any particle actually has mass.

To be a little less simplistic, the explanation offered by Peter Higgs (and various others) was that space is filled with a scalar field, called the Higgs field, and through interaction with that scalar field, particles exhibit mass. One consequence of this prediction was that certain interactions would result in the manifestation of a particle — a spin-0 boson — that came to be called the Higgs boson. The existence of the Higgs boson was confirmed on March 14, 2013, based on research at the Large Hadron Collider in CERN's complex in Geneva, Switzerland.

A THEORY OF EVERYTHING?

While quantum physics can explain many of the fundamental forces that hold particles together (or push them apart), physicists have not yet found a way to provide a quantum physics description of gravity, or a theory of *quantum gravity*. Gravity is described by Einstein's theory of relativity. Although it works great when dealing with large objects — planets, stars, and galaxies — the theory doesn't mesh with the quantum physics approach that works at the subatomic scale.

Scientists have made many attempts to unify gravity with quantum physics over the years, from Einstein's fixation on a Grand Unified Theory to the more modern approach of superstring theory. These attempted explanations represent a search for a "theory of everything" because they project the hope that quantum physics and relativity — when blended together — will offer a complete fundamental description of reality.

Many physicists who work in this area have hypothesized that a theory of quantum gravity would look similar to the theories for other fundamental forces, although dealing with large objects, on the scale of rocks, people, planets, and stars, rather than electrons, quarks, and atoms. One prediction of this is that quantum gravity, too, would be mediated by a gauge boson. This hypothetical gauge boson has been dubbed the *graviton*. This hypothetical graviton would need to have a spin of 2 and would be massless (since it would travel at the speed of light).

In recent years, scientists used the Large Hadron Collider to detect the Higgs boson and conducted astronomical observations to detect gravity waves. However, such experiments have still detected no graviton. And so, researchers have no experimental evidence for the graviton's existence nor, at the moment, even a coherent quantum gravity theory to guide the search!

Discovering How Quantum Physics Changed the World

Many topics in this book can seem abstract, so I'm pleased to reach a point where I can focus on some concrete consequences of quantum physics. As I write this book, nearly a century has passed since Werner Heisenberg gathered with some friends to break the code that would create a formal mathematical basis (matrices) for quantum mechanics. In the intervening time, the insights from quantum physics have led to technologies that people use every day — usually without thinking about how revolutionary they are.

Letting there be light with lasers

Laser technology has become ubiquitous in our world, and it definitely could not exist without an understanding of quantum physics. As with many of the insights and discoveries in this book, the development of lasers all started with Albert Einstein.

Consider an atom in an excited energy state. That excited state exists because an electron has been bumped into a higher energy level within the atom. (You can look at the energy states in the hydrogen atom in Chapter 14.) This higher energy can't last forever, though. Eventually, the electron releases that energy by dropping back down into a lower energy state. When it does so, the resulting release of energy — in the form of a photon — is a *spontaneous emission*. The light that is emitted during this process has a random phase and direction.

In 1917, Einstein predicted the possibility of a *stimulated emission*, which occurs when a photon interacts with an atom in an excited energy state and causes that atom to drop into a lower energy state. This theory predicted that researchers would find a way to stimulate the energized atoms so that they released light that was uniform in phase and direction.

Physicists conducted experiments related to the stimulated emission theory and achieved successful results with these creations.

>> **Masers:** In the early 1950s, physicists created a device for the "microwave amplification by stimulated emission of radiation," or *maser*. Because microwaves are invisible, this maser device didn't create the light show that people now expect of lasers.

>> **Lasers:** By 1958, they expanded on the principles of the maser to create an optical maser, which they cleverly renamed the *laser* (for "light amplification by stimulated emission of radiation").

REMEMBER

Lasers are a truly versatile discovery based on the principles of quantum physics and are used in many industries, ranging from information technology to surgery. The key to why a laser is so useful — and is more than just a pretty light — lies in the distinction between spontaneous emission (with random phase and direction) and stimulated emission (with uniform phase and direction). All of the light in a laser is in phase, which means that you can carefully track and quantify changes to that light. You could never use random light emissions to read a barcode at a grocery store, but when the light is all in phase, the laser can read the barcode precisely.

Controlling the flow with semiconductors and transistors

The growing understanding of matter that comes with quantum physics and the exploration of the atom has led to the ability to create whole new materials that aren't readily available in nature. Growth in the use of electricity is directly tied to our growing understanding of material science. Electricity flows easily through a *conductor*, a material such as copper wire that readily transmits electrical current. Other materials, such as wood, are poor conductors, and completely stop the flow of electricity.

But quantum physics gave researchers access to materials that don't just exist in nature (like copper and wood do). Instead, these materials enabled the researchers to build devices that provide a detailed level of control to the flow of electrons in electrical circuits. These devices include

>> **Semiconductors,** which act as good conductors in some situations but can also be modified to inhibit the flow of electrons. The special physical properties of a semiconductor allow you to turn the flow of electricity on and off through a junction or switch and, thereby, to have more control over the electrical flow.

>> **Transistors,** which apply the principle of semiconduction to tiny electronic devices. In a transistor, the flow of electricity to one set of terminals releases electrons to flow in the other direction. The transistor responds to electrical input and acts accordingly to direct the flow of current. The first transistor was built in 1947, and the physicists who built it received the 1956 Nobel Prize in Physics.

Harnessing the power of nuclear energy

Quantum physics led to the creation of the nuclear bomb in the Manhattan Project program and ultimately ushered in the nuclear age. While the nuclear bomb was a devastating manifestation of that age, the flip side brought a growing understanding of the possibilities that humanity had for drawing energy from nuclear interactions.

Energy through nuclear fission

Scientists have created successful nuclear energy reactors based primarily on *nuclear fission*, which involves taking unstable molecules (usually with a high number of protons in the nucleus) and extracting energy when those particles decay or are actively split. This process is elaborate and requires the refinement of radioactive material to be used as fuel in the nuclear reactor.

The use of nuclear fission reactors has pros and cons.

>> **Reducing reliance on fossil fuels:** As the particles in this nuclear material decay, they release energy in the form of heat. This heat, in turn, causes water to turn into steam, which creates the force necessary to turn huge turbines. So, in essence, a nuclear reactor is just a large steam turbine, where you are using nuclear decay as the fuel instead of burning fossil fuel.

>> **Production of radioactive waste products:** As the nuclear fission materials decay, you end up with radioactive fuel rods that are not generating useful amounts of energy but are still energetic enough to produce radioactive waste that needs to be carefully disposed of.

Is there a cleaner form of nuclear energy? Maybe.

Energy through nuclear fusion

A goal for the future of nuclear energy is *nuclear fusion*, which occurs when you take two lighter elements and fuse the atoms together into a single larger atom. This process is what happens inside of stars, for example, so it's known to be effective for creating heat and light — in stars, at least.

Stars can successfully sustain nuclear fusion because they have such large masses that gravity provides the power to fuse those atoms together. On Earth, humans must supply the energy to cause the nuclear fusion of the atoms. The problem is that it usually takes more energy to fuse the atoms together than you get out of the final reaction. This result makes for an unsuccessful source of energy.

Regarding taking more energy to start a fusion reaction than the reaction produces), I said "usually" because the U.S. Department of Energy announced in December 2022 that researchers at the Lawrence Livermore National Laboratory in California achieved a net gain in energy during a fusion reaction. And this experiment was replicated in the summer of 2023. If these experiments bear out, then nuclear fusion may become a viable form of energy generation in, or perhaps even sooner.

REMEMBER

The big plus for fusion over fission is in the waste product. When you use fusion to combine atoms, the end result is a heavier atom, but that atom isn't itself radioactive. If you combined hydrogen atoms together, for example, you'd get helium. Instead of disposing of radioactive waste, you would be disposing of helium — a much less risky task for the environment.

Going solar

In Chapter 3, I explore some of the earliest discoveries in quantum physics. One of them was Einstein's explanation of the photoelectric effect, where light energy caused the release of electrons from certain materials. Today, this concept is the foundation of revolutions in the creation of solar panels to extract solar energy from the renewable source of sunlight.

Of course, people didn't need quantum physics to figure out how to turn solar energy into energy that was usable. The use of sunlight for energy predates humanity, after all! Plants grow from sunlight. Our earliest form of fuel, wood for burning, grew from sunlight. Ancient plants and the ancient animals that ate them eventually died, got buried, and over the centuries turned into coal and other fossil fuels, which humans would eventually use as forms of energy to fuel our society.

But the quantum understanding of matter and energy meant that humanity was able to design and build solar panels, which skip those steps (growing, living, eating, dying, and so on) and turn solar energy directly into electricity. As the photons from the sun collide with the materials on a solar panel, they release electrons to create a flow of electrical current through the connected circuitry.

Looking Ahead at Quantum Computers

Since quantum physics tells you that the world is fundamentally uncertain, I won't foolishly try to predict the future with certainty. But quantum physics has led to major technological revolutions over the past century, and I will be a little surprised if it stops contributing. And so, I take a quick look at quantum computers, one of the new frontiers physicists and engineers are exploring.

Calculating with superposition states

A *quantum computer* is designed so that instead of regular switches that are either on or off, representing the bits of information in computer logic, you instead have *qubits* (which I introduce briefly in Chapter 4). These qubits represent electrons or possibly photons that are entangled together in quantum superposition, meaning they are represented probabilistically. Instead of being on or off for sure, a quantum bit might be in a superposition that gives it a 10 percent probability of being on and a 90 percent probability of being off.

So how does this help calculate things? In a quantum computer, the system can calculate in a superposition of possible states, meaning it can perform multiple calculations simultaneously. You could perform calculations exponentially faster with a quantum computer than you can now with a classical computer!

Sustaining and evolving quantum computers

The big problem with quantum computing is that you must keep the system entangled in a quantum superposition of states, so that the states don't collapse into one state too early. This is a delicate process because any direct interaction with any atoms or even unintended particles can break the necessary entanglement. Still, in recent years, researchers have demonstrated success with entangled quantum systems that can perform basic logic operations, so the principle is conceptually confirmed. But they're still researching effective ways to scale up these quantum computation systems to be really useful.

The possibilities of quantum computing could be amazing and far-reaching, but here I present an often-discussed application: breaking encryptions. Modern encryption largely depends on massively large numbers that are the product of two extremely large prime numbers. Modern computers are completely unable to derive the factors of these massively large numbers (find the two prime numbers), and thus, the encryption holds. But a quantum computer would be able to compute the possible factors with significantly fewer steps, and so would be able to break current ordinary encryption. Fortunately, in 2022, the U.S. Department of Commerce's National Institute of Standards and Technology (NIST) selected a variety of tools that will become part of the post-quantum encryption standard, which they are in the process of finalizing.

IN THIS CHAPTER

» Explaining why interpreting quantum physics is optional

» Describing the observer effect

» Exploring the main quantum physics interpretations

» Looking at famous criticisms of quantum physics

» Proving quantum physics with important experiments

Chapter **6**

Quantum Cats and Spooky Action: Interpretations of Quantum Physics

I f you've ever gotten into a discussion about quantum physics at a party, or if you were a fan of the television series *The Big Bang Theory*, then this chapter is likely the one that you've been waiting for. And although quantum physics works in predicting and determining the behavior of particles, a fascinating debate still rages over why and how it works.

Early experimenters in quantum physics ran into a problem: The experiments they tried didn't seem to work like any other form of experiment. Normally (meaning in classical physics), when you run an experiment, the timing on checking results doesn't particularly matter. Whether you checked the results while the experiment was ongoing or at some later date, you'd get the same results. But suddenly, in certain quantum physics setups, taking measurements at different times would result in all sorts of different outcomes and behaviors. What was happening?

In this chapter, I go over the difference in measurements and their related interpretations in the world of quantum physics. First, I spend some time reminding you that debates about interpretation or results don't actually change the results of quantum physics. Second, I explain the dominant interpretations of what is happening when you take a measurement of a quantum mechanical system. Third, I cover some of the most heated debates surrounding quantum physics, including the Schrödinger's cat thought experiment and a paradox proposed by Einstein. Finally, I look at the experiments that helped solidify the belief that quantum physicists are right to embrace quantum physics, even in light of these paradoxes.

Questioning What Needs Interpretation

Before diving into the interpretations of quantum physics, I want to cover a couple of the issues that are central to them. This approach gives me a good chance to reiterate some of the more unusual aspects of quantum physics.

Interpreting probability and measurement

As I discuss in Chapter 2, probability in classical physics represents our ignorance about the outcome of a system because we can't measure all of the factors. If I roll a 6-sided die, I can say I have a one-sixth probability that the die will land with a 6 face up; the outcome is a probability because I can't calculate all of the physical factors that affect the roll.

TIP

The role of probability in quantum physics is more fundamental than its role in classical physics. In quantum physics, the wave function, ψ, represents the probability of a particle existing in a given state or location at a given time. This probability isn't just a consequence of the physicist not having enough information but, instead, seems to be a consequence of the underlying quantum physical structure. In a sense, the particle doesn't have a set state *until it suddenly does have a set state*.

The interpretations I describe in this chapter try to explain the meaning of that quantum probability and how it ties into the act of outcome measurement.

Interpreting the observer effect

One of the key factors that interpretations of quantum physics try to solve is the *observer effect*, which is the fact that, in quantum physics, the role of taking a measurement — making an observation — has a significant impact on the outcome of the experiment.

REMEMBER

A great example of the observer effect occurs in the double slit experiment, which I describe classically in Chapter 2 and then quantum mechanically in Chapter 3. A beam of light (or a stream of particles) collides with a barrier that has two slits. When you aren't tracking which slit the light or particles (or light particles) go through at the barrier, you get a wave-based interference pattern on the other side (the screen that beam of light reaches). If you make a measurement to track which slit the particles go through, you get a different pattern, displaying no wave interference. In the sections ahead, I look at how each interpretation explains this strange behavior.

Here is where the observer effect gets even weirder: If you put a light particle detector at the slit on the double slit experiment, but don't turn it on, you still get the interference pattern. This situation means that you can't entirely attribute the observer effect to the physical interaction of the particle or photon with the detector, because the detector is still there and just not detecting anything. But the instant you turn the detector on, a measurement is being taken, and the interference pattern goes away. (See the nearby sidebar, "Conscious Observers Not Needed," for more on this phenomenon.)

TECHNICAL STUFF

CONSCIOUS OBSERVERS NOT NEEDED

A common trend, particularly among non-physicists, is to say that the observer effect requires a *conscious* observer. But there is no evidence that this is the case. The key factor is the interaction that extracts data. If a classical system interacts with the quantum mechanical system, that seems to be sufficient to cause the no-interference-pattern effect. Even if you never look at the detector's data to find out which slit a particle passed through, the mere fact that something interacted with the particle to get the information seems to be enough to destroy the interference pattern! Conscious knowledge of that information isn't necessary.

(continued)

(continued)

Furthermore, sometimes people discuss using the observer effect as proof for the existence of God. The argument usually goes something like this:

- The observer effect requires an observer.

- The universe existed before any observers did (except for possibly God).

- God would be an observer of all things.

- Therefore, God must exist, or else the wave function would never have collapsed.

As a logical argument, let alone a scientific one, this line of thought fails on a few counts. To start with, the first point in the argument represents a misunderstanding of what quantum physics says about the observer effect. The wave function would have collapsed shortly after the universe sprang into existence — as soon as particles began interacting with multiple other particles. (See the section "Outlining Three Quantum Physics Interpretations" later in the chapter for information about wave functions.)

More importantly, though, even if you assume the first point in the argument as valid, the third point creates a contradiction. The observer effect says that when an observation is made, event A happens, and when there is no observation, event B happens. But if the third point is true, then there's *always* an observation being made. Physicists would *always* see event A and would never see event B at all. The fact that the observer effect exists at all suggests that (if a deity exists) the deity's either not peeking at quantum mechanical events or that the peeking doesn't count as a physical observation.

Entanglement revisited

Another important element in quantum physics interpretations, and their criticisms, is the idea of entanglement, which I first introduce in Chapter 1. This idea says that when parts of a system interact, those pieces of the system become engaged together, or entangled, in an important way. Their states are tied together, so that understanding a state of one part of the system means you have information about other parts of the system.

On its surface, this situation doesn't seem all that unusual. In Chapter 1, I give an example with pennies that is perfectly easy to understand in classical terms. But when you consider how probability applies in quantum physics (with states not being only probabilities until measurements are made), you have a problem. The discussion of quantum theory criticisms (in the section "Enduring Debates, Bickering, and Other Counterarguments" later in this chapter) focuses precisely on the probability problem in two different, but important, ways — the Schrödinger cat problem points out the absurdity of entanglement of macroscopic objects and the EPR paradox focuses on the difficulty of how the entanglement could be communicated.

Discovering That the Most Common Interpretation Is to Shrug

Before I begin discussing deep questions on the interpretations of quantum physics, I have to spend a moment reiterating a key point: For most physicists working on quantum physics, the questions of interpretation have no bearing on their day-to-day work on the subject.

Some quantum physicists have gone so far as to say that interpretations are nothing more than a distraction from the actual work of doing quantum physics. Parts 3 and 4 involve digging deeply into the mathematics of quantum mechanical equations and learning how to apply them in particular ways.

Just be aware of these two points:

>> None of the physicists who disagree about the conceptual understanding of quantum mechanics actually disagree about how to solve the equations.

>> The results of the equations give you values that match with the growing number of experiments validating quantum mechanics.

REMEMBER

In a typical quantum mechanics class, you briefly cover the conceptual matters and the possible interpretations, and then move forward into the mathematical complexities. If you're studying for a degree in physics, the thing that matters about quantum physics is how you solve the equations. The debate about the underlying nature of reality, while interesting, is a secondary consideration.

Outlining Three Quantum Physics Interpretations

I focus on three main interpretations of quantum physics in this chapter. Physicists built these interpretations with the understanding that the equations of quantum mechanics work and represent actual physical behaviors. The physicists who formulated these interpretations did so to figure out explanations for the observed physical behaviors of matter and energy by answering this fundamental question:

What happens physically when a quantum mechanical measurement is made and the system goes from a wave of probabilities to a definite outcome?

REMEMBER

A quantum mechanical system is represented at one point as a probabilistic wave function (ψ), and then — when a measurement is made — that probabilistic wave function *seems to become* a specific, definite value. This understanding works to fine to solve the equations and explain the experiments, but the lack of satisfaction with it as a conceptual explanation is why a debate over the meaning of quantum physics still lingers a century after its development.

The popular one: Copenhagen interpretation

The *Copenhagen interpretation* is so named because it was the one strongly advocated by Niels Bohr and his Copenhagen Institute. It is also the first firmly established interpretation of quantum physics. In this view of quantum physics, the moment a measurement is made, the wave function *collapses* into a definite value. The act of making the measurement causes this collapse.

Consider the case of the observer effect in the double slit experiment (refer to the section "Interpreting the observer effect," earlier in the chapter, for a description). The following statements sum up how the Copenhagen interpretation looks at what happens during the experiment.

>> **When you make a measurement at a slit:** The wave function collapses at the slit *before* it can undergo any interference. A definite particle moves from the slit to the screen. No interference pattern shows up.

>> **When you make a measurement at the screen:** The wave function passes through the two slits, resulting in two wave functions. The two wave functions interfere with each other and then collapse at the screen, showing the pattern of that interference.

The problem with this approach is that it really just describes the collapse of the wave function as an event that happens, without an actual physical description of what is happening. Why does the act of measuring cause the wave function to collapse? Physicists don't know. That's just what happens, says the Copenhagen interpretation. For those who hold onto this interpretation, that explanation is enough.

The fun one: Many worlds interpretation

In the *many worlds interpretation*, an untold number of universes exist, and each one represents one possible branch of quantum probability. All possible outcomes of the event occur when a measurement happens, but they occur in different versions of the universe (or different "worlds"). You only see one of those outcomes, because the "you" that looks at the result of the measurement is in just one of the worlds. Somewhere — in a different version of the universe — a different version of you sees the other possible outcomes.

This no doubt sounds like the stuff of science fiction, in part because so much science fiction has been based on this premise. The stories usually involve someone getting in contact with one of the other worlds, which represents one of the other probability branches. But don't be fooled. Within the many worlds interpretation of quantum physics, this science fiction scenario is completely impossible. The interpretation assumes that once the branching happens, the split worlds will never reunite or interact in any way. They will continue to change and evolve completely independently of each other.

American physicist Hugh Everett III proposed the many worlds interpretation in 1957. The goal was to take the multiple outcomes of the wave function seriously. In a very real sense, in this interpretation the wave function never collapses. The wave function of the entire universe keeps evolving and growing superposed with more states, and more possible states of the universe, every time there's a quantum event. Observers don't see the multiple outcomes because they see the outcome for only one state of the universe — the one that they're currently in. The result of observing the definite outcome looks like a wave function collapse is happening.

The benefit of this interpretation removes the need to explain the collapse of the wave function, but its flaw is probably pretty obvious: You have to assume the existence of untold other worlds that are completely inaccessible to us. That's a pretty steep price to pay.

The responsible one: Hidden variables

Another approach to explaining quantum mechanics that has a bit of a following is known as the *hidden variables interpretation*. It's also sometimes called the *de Broglie-Bohm interpretation* after a concept originally put forward by Louis de Broglie and then expanded upon by David Bohm. Most physicists would *like* for this interpretation to be true, but it probably isn't.

The hidden variables interpretation resolves the issues of probability and uncertainty in quantum physics by saying that, just like in classical physics, the issues are merely a measure of humanity's ignorance. This interpretation says that some force guides the quantum processes, but physicists and researchers just don't have the knowledge about how to detect or fully account for it. In de Broglie's original concept, he called this force a *pilot wave*, although Bohm's later work referred to it as a *quantum potential*.

TIP

To be consistent with the results of quantum physics experimentation, this proposed quantum potential has to be *nonlocal*. That is, it must have information about all of the other particles that a specific particle is entangled with — which includes any particle that it's ever interacted with — and that information must be instantaneously available.

The benefit of this interpretation is that it restores logical consistency and determinism to the quantum universe because probability and uncertainty go away. The flaw is that the approach just *asserts* that logical consistency and determinism exist and manifest in the form of a *completely undetected hypothetical quantum potential*. The potential is not only undetected but would also need to represent everything about the universe in order to function. At that point, you basically have a wave function that represents the entire universe. And due to this inconceivable complexity, the approach isn't adopted by many physicists.

Enduring Debates, Bickering, and Other Counterarguments

Most of the debates about quantum physics center on the Copenhagen interpretation (discussed in the section "The popular one: Copenhagen interpretation," earlier in this chapter), because that's historically been the most widely held interpretation. In that interpretation, the probabilistic wave function collapses when a measurement is made and transitions from a general description of the probability of particles being in different states to a distinct outcome of a particle in a particular state. The most famous criticisms of that approach came from two of the most important physicists of the twentieth century: Erwin Schrödinger and Albert Einstein. Both proposed thought experiments intended to prove the absurdity of the Copenhagen interpretation.

Addressing the (dead?) cat in the room

Erwin Schrödinger proposed the classic Schrödinger's cat thought experiment in 1935. *Note:* No real cats were harmed in the making of this thought experiment!

The Copenhagen interpretation indicates that the wave function collapses into a final state when a measurement is made. So, if you had a radioactive isotope that had a 50 percent chance of decaying in the next hour, and it was inside a sealed box, you would say that the wave function of the isotope existed in a superposition state of both decayed and not decayed until a measurement was made.

This is crucial: It's *not* just that a physicist doesn't know if the isotope has decayed, it's that according to the Copenhagen interpretation the isotope's state is genuinely undetermined until the box is opened and the measurement is made. The isotope is considered in a superposition of quantum probabilities — where it is both decayed and not-decayed at the same time — until the box is opened and the wave function collapse happens.

Schrödinger thought this was complete nonsense and proposed a scenario to prove it. Assume you have exactly the same box, but now it contains a device connected to a Geiger counter. When the radioactive isotope decays, it triggers the Geiger counter, which will then trigger the device in the box to break a bottle of poison gas. Oh, yeah, and there's also a cat in the box, so if the isotope decays, the cat will instantly die. (Again, Schrödinger was proposing this as only a thought experiment.)

According to the Copenhagen interpretation of quantum physics, until the point that the box is opened, the isotope is in a superposition state of being both decayed and not decayed, which means that the cat is in a state where its wave function exists in a superposition of both alive and dead. But once the box is opened, a measurement is made (you can see inside it), and the wave function collapses into a state where the cat is either alive or dead.

In a physical sense, the explanation around the experiment notes that a measurement takes place the moment the isotope decays. The Geiger counter, the device, the bottle of poison gas, and the cat all instantly react appropriately based on that measurement. Alternatively, if the isotope doesn't decay, the Geiger counter knows that as well, remains inactive, and the cat accordingly remains alive. So the thought experiment is absurd precisely because the quantum system is interacting with a classical system inside the box, so the collapse does actually happen in this case well before the box is opened.

Although the Schrödinger's cat thought experiment ultimately failed to show that the Copenhagen interpretation was absurd, it did become a useful way of:

>> **Highlighting the logical inconsistencies of the quantum world** when expanding it into the macroscopic world.

>> **Implying the existence of a boundary to quantum logic.** Somewhere between the decay of the particle and the death of the cat, human intuitions tell you that a boundary must exist where the quantum logic fails. Unfortunately, physicists are unable to define precisely where that boundary is.

The other quantum physics interpretations deal with this thought experiment differently than does the Copenhagen interpretation:

>> **In the many worlds interpretation,** the universe splits into multiple universes. In some of those universes, the cat will be alive and in some the cat will be dead. You just don't know which type of universe you're in until you make the measurement.

>> **In the hidden variables interpretation,** some undetected quantum force guides the collapse of the wave function and determines whether the cat lives or dies. But the experiment's observers don't know what the force is and can't measure it, and the probability is just a result of this ignorance.

Getting spooked with Einstein

One conceptually troubling aspect of the quantum mechanics wave function is the idea of two particles becoming *entangled*. Basically, anytime an interaction occurs that would determine or change a particle's state, you have a possibility of the particles involved becoming entangled together. This feature of quantum physics led to one of Einstein's most formal criticisms of quantum mechanics.

Einstein — together with Boris Podolsky and Nathan Rosen — formulated his criticism in 1935 (the same year the Schrödinger's cat thought experiment came about). The criticism became known as the Einstein-Podolsky-Rosen paradox, or the *EPR paradox*, although it was mainly written by Podolsky. A more straightforward and popular version of the EPR paradox was proposed by David Bohm (of the hidden variables interpretation) in 1951, so that's the version that I describe in this section.

Setting assumptions and relating theories

Assume you have an experiment that emits an electron and positron pair. These two particles are created so that they are entangled with each other. If you measure one and identify it as an electron, then you instantly know that the opposite particle is a positron, and vice versa. The particles in this thought experiment follow these steps:

1. **The electron and positron are created and go flying off in different directions, toward detectors (A and B) set up a far distance away.**

2. **When a particle reaches its respective detector, it is measured (identified) as either an electron or positron.**

   Each detector has a 50 percent chance of finding an electron and also a 50 percent chance of finding a positron.

3. **As guided by the Copenhagen interpretation, as soon as one detector measures (identifies) its particle, the wave function collapses down into a consistent state.**

If detector A measures an electron, then detector B will always measure a positron.

The contradictory part of the EPR paradox is in reconciling Steps 2 and 3 in the preceding step list: How does the information get from the identified state of the particle at detector A to detector B fast enough to cause an instantaneous collapse?

This problem with reconciling an instantaneous collapse comes from Einstein's theory of relativity, which has an important consequence: The speed of light seems to be the fastest that anything can move. How, in the EPR paradox scenario, does the quantum message get from the particle at detector A to the particle at detector B instantaneously? It seems that this setup results in some sort of information being exchanged between the particles at a speed faster than the speed of light. Einstein referred to this sort of behavior as "spooky action at a distance," and it was one of the things about quantum physics that perpetually bugged him.

Looking for a complete interpretation

The goal of the EPR paradox wasn't to disprove quantum physics. It clearly works, and nothing in the thought experiment implies that it doesn't. What the criticism suggests, however, is that quantum physics interpretations appear to be an *incomplete* description of reality. Einstein (and many others) were looking for a complete description of reality. (See Chapter 5 for more on this search.)

As you find out in the next section, experiments have confirmed that quantum physics is right about what happens in this EPR paradox scenario. A measurement of one of these particles does immediately impact the other particle. These two entangled particles are definitely displaying a nonlocal relationship to each other.

Experiments that show an immediate impact between measured particles don't necessarily imply that any signal is being sent between the two particles. The particles are entangled, after all, which means the wave function that describes them is in two states. Either it's "electron at detector A, positron at detector B" or "positron at detector A, electron at detector B." A measurement made at either detector is enough to collapse the wave function into the correct state. Thinking that information must be communicated across space is a misunderstanding, according to the current understanding of the Copenhagen interpretation.

Exploring Entangled Experiments

While the debates around quantum physics continued, physicists explored ways to test the different interpretations. The questions about entanglement in the Schrödinger's cat and EPR paradox examples were intended as thought experiments, but physicists also wanted a way to actually test entanglement in the real world.

Bell's inequality

In 1964, Irish physicist John Bell took a year off from his normal research. What did he do? He figured out how to test the EPR paradox. What else would someone do on their vacation?

Bell started his research with these thoughts and assumptions:

>> Bell considered what it would mean if the Copenhagen interpretation was wrong and instead, some sort of hidden variable guided the outcome of the measurements.

>> To preserve relativity, Bell assumed that Einstein's understanding of the speed of light was valid. So, any hidden variables would have to communicate at less than the speed of light, and they wouldn't exhibit nonlocality (see the section "The responsible one: Hidden variables," earlier in the chapter for info about nonlocality). They'd be *local hidden variables*.

>> Bell proposed to experiment with two entangled particles that had opposite spin. One of the particles had up spin, and one had down spin. An experimental setup called Stern-Gerlach magnets can detect this spin (as described in Chapter 11).

What Bell showed was that for certain configurations of these two sets of magnets, the probabilistic Copenhagen interpretation and the local hidden variable interpretation yielded different results. The mathematical relationship that described the expected outcome with local hidden variables is called *Bell's inequality*. In the standard probability approach matching the Copenhagen interpretation, the results would *violate* Bell's inequality.

The great thing about a result like this is that it gives experimental physicists something to do. Here you have a real foundational concept that finally has a way to test it! Bell already did all of the hard math, after all. When would physicists be able to put these concepts to the test?

Alain Aspect's grand experiment

The French physicist Alain Aspect would finally put Bell's inequality to a test in 1981 by building on the 1972 efforts of American physicist John Clauser. Clauser's experiments showed a violation of Bell's inequality, but his equipment and experimental setup weren't quite so refined enough to be conclusive. Physicists were able to pick holes due to the possible flaws in the 1972 experiment. It took another decade of technological and experimental progress to create an experiment that would convince everyone.

One of the concerns was that the local hidden variables were somehow tied to the specific experimental configuration. In other words, the particles in essence "knew" (via the local hidden variables) where they were going, and could orient accordingly. Aspect devised some clever tricks to avoid this problem, so that the path and detectors could be changed in mid-experiment.

REMEMBER

When Aspect conducted and published the results of his experiments, the outcome was pretty conclusive: Bell's inequality was violated. There was no room in quantum physics for local hidden variables. The probabilistic interpretation of quantum physics had held up to its most direct experimental test.

Forty years after Aspect published his results, Aspect and Clauser shared the 2022 Nobel Prize in Physics for their research (together with Austrian physicist Anton Zeilinger, who also did groundbreaking research with entangled particles, a key element of the quantum computer research discussed in Chapter 5). John Bell, who had also contributed to this work, had died of a cerebral hemorrhage in 1990, and the Nobel Prize isn't awarded posthumously. Still, the recognition that Aspect and Clauser received for this amazing experimental achievement also signifies a recognition of how important Bell's contribution was to the fundamental understanding of quantum physics.

3

By the Numbers: Basic Quantum Physics Math

Get familiar with writing quantum physics equations for wave functions.

Find wave functions for particles in different energy wells.

Discover wave functions for quantum harmonic oscillators.

Apply quantum physics to angular momentum.

Get deeper into the mathematics of particle quantum spin.

IN THIS CHAPTER

» **Creating state vectors**

» **Using Dirac notation for state vectors**

» **Working with bras and kets**

» **Understanding matrix mechanics**

» **Getting to wave mechanics**

Chapter **7**

Entering the Matrix: Welcome to State Vectors

Quantum physics isn't just about playing around with your particle accelerator while trying not to destroy the universe. Sometimes, you get to do things that are a little more mundane, such as turning lights off and on, performing a bit of calculus, or playing with dice. If you're actually exploring aspects of physics with your dice, the lab director won't even get mad at you. You can assemble the probability values into a vector (single-column matrix) in *Hilbert space* (a type of infinitely dimensional vector space with some properties that are especially valuable in quantum physics).

Much of this chapter builds on established understanding for anyone who's had a course in linear algebra and is comfortable with vectors, though it may be a very different approach to vectors than you've seen before. If you're not previously comfortable with vectors, a quick break to check out *Linear Algebra For Dummies* might be in order before diving into this chapter.

This chapter introduces how you deal with probabilities in quantum physics, starting by viewing the various possible states a particle can occupy as a vector — a vector of probability states. From there, I help familiarize you with some mathematical notations common in quantum physics, including bras, kets, matrices, and wave functions. Along the way, you also get to work with some important operators.

Creating Your Own Vectors in Hilbert Space

Suppose that you've been experimenting with rolling a pair of dice and are trying to figure the *relative probability* (the degree of confidence that you may have for the occurrence of a certain result) that the dice will show various values. (You may want to peek at Chapter 2 where I address classical probability.) You come up with a list indicating the relative probability of rolling a 2, 3, 4, and so on, all the way up to 12. A table of these relative probabilities looks like the following.

Sum of the Dice	Relative Probability (Number of Ways of Rolling a Particular Total)
2	1
3	2
4	3
5	4
6	5
7	6
8	5
9	4
10	3
11	2
12	1

The table tells you that you're twice as likely to roll a 3 than a 2, four times as likely to roll a 5 than a 2, and so on. You can assemble these relative probabilities into a vector (if you're thinking of a traditional "vector" from physics, think in terms of a column of the vector's components, not a magnitude and direction) to keep track of them easily. You denote the vector as follows:

$$\begin{bmatrix} 1 \\ 2 \\ 3 \\ 4 \\ 5 \\ 6 \\ 5 \\ 4 \\ 3 \\ 2 \\ 1 \end{bmatrix}$$

Okay, now you're getting closer to the way quantum physics works. You have a vector of probabilities that the dice will occupy various states (that is, have certain values). However, quantum physics doesn't deal directly with probabilities but rather with probability amplitudes, which are the square roots of probabilities. To find the actual probability that a particle (in this case, the dice) will be in a certain state, you add wave functions — which are going to be represented by these vectors — and then square them (see the discussion in Chapter 3). So, you take the square root of all these entries to get the probability amplitudes:

$$\begin{bmatrix} \sqrt{1} \\ \sqrt{2} \\ \sqrt{3} \\ \sqrt{4} \\ \sqrt{5} \\ \sqrt{6} \\ \sqrt{5} \\ \sqrt{4} \\ \sqrt{3} \\ \sqrt{2} \\ \sqrt{1} \end{bmatrix}$$

The vector of square roots better represents the total probability, and adding the squares of all these entries should add up to a total probability of 1 (because at least one of these outcomes will take place). As it is now, the sum of the squares of these numbers is 36, so you must divide each entry by $36^{1/2}$, or 6, and the vector now looks like this:

$$\begin{bmatrix} \sqrt{1/36} \\ \sqrt{2/36} \\ \sqrt{3/36} \\ \sqrt{4/36} \\ \sqrt{5/36} \\ \sqrt{6/36} \\ \sqrt{5/36} \\ \sqrt{4/36} \\ \sqrt{3/36} \\ \sqrt{2/36} \\ \sqrt{1/36} \end{bmatrix}$$

So now you can get the probability amplitude of rolling any combination from 2 to 12 by reading down the vector; the probability amplitude of rolling a 2 is $\frac{1}{6}$, of rolling a 3 is $\frac{\sqrt{2}}{6}$, and so on.

Making Life Easier with Dirac Notation

This new state vector (see the preceding section) can be thought of as a vector in dice space — all the possible states that a pair of dice can take, which is an 11-dimensional space.

But in most quantum physics problems, the vectors can be infinitely large — for example, a moving particle can be in an infinite number of states. Handling large arrays of states isn't easy using vector notation, so instead of explicitly writing out the whole vector each time, quantum physics usually uses the notation developed by physicist Paul Dirac — the *Dirac notation* or *bra-ket notation*.

I tell you the meaning of these terms in a bit, but right now you're probably thinking, "How do I pronounce that?" To end the suspense, I can tell you that you pronounce it just like it's spelled. The "bra" (yes, just like the undergarment) rhymes with "rah." The "ket" rhymes with "get." So, if you're saying "bra-ket" and getting something that sounds roughly like "rocket" with a "b" at the beginning, you're pretty close. (Google tells me that about 112 pronunciations of *Dirac* exist in British English, so that suspense must remain.)

Abbreviating state vectors as kets

Dirac notation abbreviates the state vector as a ket, which is the name for a vector that uses this notation: $|\psi\rangle$. So, in the dice example from the previous section, you can write the state vector as a ket in this way:

$$|\psi\rangle = \begin{bmatrix} 1/6 \\ \sqrt{2}/6 \\ \sqrt{3}/6 \\ 2/6 \\ \sqrt{5}/6 \\ \sqrt{6}/6 \\ \sqrt{5}/6 \\ 2/6 \\ \sqrt{3}/6 \\ \sqrt{2}/6 \\ 1/6 \end{bmatrix}$$

Here, the components of the state vector are represented by numbers in 11-dimensional dice space. More commonly, however, each component represents a function of position, x, and time, t. But this vector doesn't represent anything this elaborate (yet). Baby steps. Quantum baby steps.

TECHNICAL STUFF

THE BASIS OF VECTORS

Think for a minute about vectors in three dimensions. These vectors have three coordinates, corresponding to three dimensional axes: x, y, and z. These three axes can each be treated as its own vector, an x-vector, a y-vector, and a z-vector. You can't write an x-vector as a *linear combination* of y-vectors and z-vectors. The three axis vectors are *linearly independent* of each other.

Similarly, a set of vectors in Hilbert space is *linearly independent* if you can't write any of the vectors as a linear combination of the others. These linearly independent vectors can be treated as independent axes and be used to form a valid basis in Hilbert space. In other words, you'll be able to write any state vector as a linear combination of these basis vectors, the same way you can write a three-dimensional vector as a linear combination of an x-vector, a y-vector, and a z-vector.

One of the benefits of the bra-ket notation is that it avoids the need to determine a full basis, as described in the section "Covering all your bases: Bras and kets as basis-less state vectors," later in the chapter.

Writing the Hermitian conjugate as a bra

For every ket, there's a corresponding bra. (The terms come from bra-ket, or bracket, which I explain in the section "Grooving with Operators," later in the chapter.) A bra is the Hermitian conjugate of the corresponding ket. To find this, you take the original ket, transpose it, and then take the complex conjugate. The *complex conjugate* flips the sign connecting the real and imaginary parts of a complex number. But in the case of the ket for dice space, the values are all real numbers, so the corresponding bra is this row vector:

$$\langle\psi| = \left[\frac{1}{6} \ \frac{\sqrt{2}}{6} \ \frac{\sqrt{3}}{6} \ \frac{2}{6} \ \frac{\sqrt{5}}{6} \ \frac{\sqrt{6}}{6} \ \frac{\sqrt{5}}{6} \ \frac{2}{6} \ \frac{\sqrt{3}}{6} \ \frac{\sqrt{2}}{6} \ \frac{1}{6} \right]$$

REMEMBER

Note that if any of the elements of the ket are complex numbers, you have to take their complex conjugate when creating the associated bra. For instance, if your complex number in the ket is $a + bi$ (where bi is an imaginary number), its complex conjugate flips to the opposite operator so that in the bra, it is $a - bi$, and vice versa.

Multiplying bras and kets: A probability of 1

You can take a bra and a ket, and combine them together, $\langle\psi|\psi\rangle$, which means taking the product of them, like this:

$$\langle\psi|\psi\rangle = \left[\frac{1}{6} \ \frac{\sqrt{2}}{6} \ \frac{\sqrt{3}}{6} \ \frac{2}{6} \ \frac{\sqrt{5}}{6} \ \frac{\sqrt{6}}{6} \ \frac{\sqrt{5}}{6} \ \frac{2}{6} \ \frac{\sqrt{3}}{6} \ \frac{\sqrt{2}}{6} \ \frac{1}{6} \right] \begin{bmatrix} 1/6 \\ \sqrt{2}/6 \\ \sqrt{3}/6 \\ 2/6 \\ \sqrt{5}/6 \\ \sqrt{6}/6 \\ \sqrt{5}/6 \\ 2/6 \\ \sqrt{3}/6 \\ \sqrt{2}/6 \\ 1/6 \end{bmatrix}$$

This calculation is just matrix multiplication, and the result is the same as taking the sum of the squares of the elements:

$$\langle\psi|\psi\rangle = \left[\frac{1}{36} + \frac{2}{36} + \frac{3}{36} + \frac{4}{36} + \frac{5}{36} + \frac{6}{36} + \frac{5}{36} + \frac{4}{36} + \frac{3}{36} + \frac{2}{36} + \frac{1}{36} \right] = 1$$

This calculation comes out the way it should because the total probability should add up to 1. Therefore, in general, the product of the corresponding bra and ket equals 1:

$$\langle \psi | \psi \rangle = 1$$

If this relation holds and produces a value of 1, the vector is normalized.

Covering all your bases: Bras and kets as basis-less state vectors

The reason ket notation, $|\psi\rangle$, is so popular in quantum physics is that it allows you to work with state vectors in a *basis-free* way. In other words, you're not stuck in the position basis, the momentum basis, or the energy basis. That's helpful, because most of the work in quantum physics takes place in abstract calculations, and you don't want to have to drag all the components of state vectors through those calculations. Often you can't because infinite possible states may exist in the problem you're dealing with.

For example, suppose that you're representing your states

>> **Using position vectors in a three-dimensional Hilbert space.** In this case, you have *x*, *y*, and *z* axes, forming a position basis for your space. That's fine, but not all your calculations have to be done using that position basis. Because position is the default way of talking about a three-dimensional space, you don't often need a special name for this, but it would be called the *position space* if you were trying to be specific.

>> **Using momentum vectors in a three-dimensional space**, with three axes in Hilbert space, p_x, p_y, and p_z. To accommodate this momentum basis, you'd have to change all your position vectors to momentum vectors, adjusting each component, and keep track of what happens to every component through all your calculations. The three-dimensional space defined with momentum vectors using the momentum basis is called the *momentum space* (to differentiate it from the position space).

So, Dirac's bra–ket notation comes to the rescue here; you use it to perform all the math and then plug in the various components of your state vectors as needed at the end. That is, you can perform your calculations in purely symbolic terms, without being tied to a basis.

And when you need to deal with the components of a ket, such as when you want to get physical answers, you can also convert kets to a different basis. Generally, when you have a vector $|\psi\rangle$, you can express it as a sum over N basis vectors, $|\phi_i\rangle$, like so:

$$|\psi\rangle = \sum_{i=1}^{N} |\phi_i\rangle\langle\phi_i|\psi\rangle$$

where N is the dimension of the Hilbert space, and i is an integer that labels the basis vectors.

Understanding some relationships by using kets

Ket notation makes the math easier than it is in matrix form because you can take advantage of a few mathematical relationships. For example, here's the so-called *Schwarz inequality* for state vectors:

$$|\langle\psi|\phi\rangle|^2 \leq \langle\psi|\psi\rangle\langle\phi|\phi\rangle$$

This inequality says that the square of the *modulus* (which is basically the equivalent of the absolute value, but when talking about vectors) of the product of two state vectors is less than or equal to the product of their individual moduli. This turns out to be another way of writing the vector inequality:

$$|A \cdot B|^2 \leq |A|^2|B|^2$$

REMEMBER

So why is the Schwarz inequality so useful? You can use the Schwarz inequality to derive the Heisenberg uncertainty principle (introduced in Chapter 3). And this principle is one of the central concepts in quantum physics because it quantifies the relationship between position and momentum. I derive it in the section "Starting from Scratch and Ending Up with Heisenberg," later in this chapter.

Other ket relationships can also simplify your calculations. For instance, two kets, $|\psi\rangle$ and $|\phi\rangle$, are said to be orthogonal if

$$\langle\psi|\phi\rangle = 0$$

TIP

Three-dimensional vectors are *orthogonal* if they're perpendicular to each other. This also means they are *linearly independent*. You can extend these concepts of orthogonality and linear independence to higher dimensions, but the metaphor with perpendicularity helps you to think about the relationships between these vectors.

And two kets are said to be *orthonormal* if they are both orthogonal to each other and if each one is also *normalized* (produces a value of 1). In other words, they must meet the following conditions:

» $\langle \psi | \phi \rangle = 0$

» $\langle \psi | \psi \rangle = 1$

» $\langle \phi | \phi \rangle = 1$

Note: Keep these relationships that use kets in mind when you work with opera-tors. See the next section.

Grooving with Operators

What about all the calculations that you're supposed to be able to perform with kets? Taking the product of a bra and a ket, $\langle \psi | \phi \rangle$, is fine as far as it goes, but sup-pose that you want to extract some physical quantities from the calculations that you can measure. That's where operators come in.

Hello, operator: How operators work

Here's the general definition of an operator (in this case, A) in quantum physics: An operator is a mathematical rule that, when operating on a ket, $|\psi\rangle$, transforms that ket into a new ket, $|\psi'\rangle$, in the same space. (This new ket could just be the old ket multiplied by a *scalar*, which is a quantity that has only magnitude and not direction). So, when you have an operator A, it transforms kets and bras like this:

$$A|\psi\rangle = |\psi'\rangle$$

$$\langle \psi | A^\dagger = \langle \psi'|$$

REMEMBER

Here are several examples of the kinds of operators you'll see in quantum mechan-ics calculations:

» **Hamiltonian (H):** Applying the Hamiltonian operator (which looks different for every different physical situation) gives you E, the energy of the particle represented by the ket. E is a scalar quantity:

$$H|\psi\rangle = E|\psi\rangle$$

» **Unity or identity (I):** The unity or identity operator leaves kets unchanged:

$$I|\psi\rangle = |\psi\rangle$$

>> **Gradient (∇):** The gradient operator (first introduced in Chapter 2) employs *derivatives* (the varying rate of change of a function with respect to the different position variables) and works like this:

$$\nabla |\psi\rangle = \frac{\partial}{\partial x}|\psi\rangle \boldsymbol{i} + \frac{\partial}{\partial y}|\psi\rangle \boldsymbol{j} + \frac{\partial}{\partial z}|\psi\rangle \boldsymbol{k}$$

>> **Linear momentum (P):** The linear momentum operator looks like this in quantum mechanics and is used to find the linear momentum:

$$P|\psi\rangle = -i\hbar \nabla |\psi\rangle$$

>> **Laplacian (∇²):** You use the Laplacian operator, which is much like a second-order gradient, to create the energy-finding Hamiltonian operator:

$$\nabla^2 |\psi\rangle = \nabla \cdot \nabla |\psi\rangle = \frac{\partial^2}{\partial x^2}|\psi\rangle + \frac{\partial^2}{\partial y^2}|\psi\rangle + \frac{\partial^2}{\partial z^2}|\psi\rangle$$

WARNING

In general, multiplying operators together is not independent of order (that is, it does not follow the commutative property, as it does in arithmetic), so for the operators A and B, AB ≠ BA.

I expected that: Finding expectation values

Given that everything in quantum physics is done in terms of probabilities, making predictions becomes very important. And the most important such prediction is the expectation value. The *expectation value* of an operator is the average value that you would measure if you performed the measurement many times. For example, the expectation value of the Hamiltonian operator (see the preceding section) is the average energy of the system you're studying.

REMEMBER

The expectation value is a weighted average of the probabilities of the system's being in its various possible states. Here's how you find the expectation value of an operator A:

Expectation value = $\langle \psi | A | \psi \rangle$

Note that because you can express $\langle \psi |$ as a row vector and $|\psi\rangle$ as a column vector, you can express the operator A as a square matrix.

For example, suppose you're working with a pair of dice and the probabilities of all the possible sums (see the earlier section "Creating Your Own Vectors in Hilbert Space"). In this dice example, the expectation value is a sum of terms, and each term is a value that the dice display when rolled, multiplied by the probability that that value will appear.

The bra and ket will handle the probabilities, so it's up to the operator that you create for this — call it the *Roll operator*, R — to store the dice values (2 through 12) for each probability. Therefore, the operator R looks like this:

$$
R = \begin{bmatrix}
2 & 0 & 0 & 0 & 0 & 0 & 0 & 0 & 0 & 0 & 0 \\
0 & 3 & 0 & 0 & 0 & 0 & 0 & 0 & 0 & 0 & 0 \\
0 & 0 & 4 & 0 & 0 & 0 & 0 & 0 & 0 & 0 & 0 \\
0 & 0 & 0 & 5 & 0 & 0 & 0 & 0 & 0 & 0 & 0 \\
0 & 0 & 0 & 0 & 6 & 0 & 0 & 0 & 0 & 0 & 0 \\
0 & 0 & 0 & 0 & 0 & 7 & 0 & 0 & 0 & 0 & 0 \\
0 & 0 & 0 & 0 & 0 & 0 & 8 & 0 & 0 & 0 & 0 \\
0 & 0 & 0 & 0 & 0 & 0 & 0 & 9 & 0 & 0 & 0 \\
0 & 0 & 0 & 0 & 0 & 0 & 0 & 0 & 10 & 0 & 0 \\
0 & 0 & 0 & 0 & 0 & 0 & 0 & 0 & 0 & 11 & 0 \\
0 & 0 & 0 & 0 & 0 & 0 & 0 & 0 & 0 & 0 & 12
\end{bmatrix}
$$

So, to find the expectation value of R, you need to calculate $\langle \psi | R | \psi \rangle$. Spelling that out in terms of components gives you the following:

$$
\langle \psi | R | \psi \rangle = \begin{bmatrix} \dfrac{1}{6} & \dfrac{\sqrt{2}}{6} & \dfrac{\sqrt{3}}{6} & \dfrac{2}{6} & \dfrac{\sqrt{5}}{6} & \dfrac{\sqrt{6}}{6} & \dfrac{\sqrt{5}}{6} & \dfrac{2}{6} & \dfrac{\sqrt{3}}{6} & \dfrac{\sqrt{2}}{6} & \dfrac{1}{6} \end{bmatrix}
$$

$$
\begin{bmatrix}
2 & 0 & 0 & 0 & 0 & 0 & 0 & 0 & 0 & 0 & 0 \\
0 & 3 & 0 & 0 & 0 & 0 & 0 & 0 & 0 & 0 & 0 \\
0 & 0 & 4 & 0 & 0 & 0 & 0 & 0 & 0 & 0 & 0 \\
0 & 0 & 0 & 5 & 0 & 0 & 0 & 0 & 0 & 0 & 0 \\
0 & 0 & 0 & 0 & 6 & 0 & 0 & 0 & 0 & 0 & 0 \\
0 & 0 & 0 & 0 & 0 & 7 & 0 & 0 & 0 & 0 & 0 \\
0 & 0 & 0 & 0 & 0 & 0 & 8 & 0 & 0 & 0 & 0 \\
0 & 0 & 0 & 0 & 0 & 0 & 0 & 9 & 0 & 0 & 0 \\
0 & 0 & 0 & 0 & 0 & 0 & 0 & 0 & 10 & 0 & 0 \\
0 & 0 & 0 & 0 & 0 & 0 & 0 & 0 & 0 & 11 & 0 \\
0 & 0 & 0 & 0 & 0 & 0 & 0 & 0 & 0 & 0 & 12
\end{bmatrix}
\begin{bmatrix}
1/6 \\ \sqrt{2}/6 \\ \sqrt{3}/6 \\ 2/6 \\ \sqrt{5}/6 \\ \sqrt{6}/6 \\ \sqrt{5}/6 \\ 2/6 \\ \sqrt{3}/6 \\ \sqrt{2}/6 \\ 1/6
\end{bmatrix}
$$

Doing the math, you get

$$\langle \psi | R | \psi \rangle = 7$$

So, the expectation value of a roll of the dice is 7. Fortunately, this matches the more conventional means of calculating the expected value for rolling two dice in a probability class. (I cover a more classical approach to this in Chapter 2.) Anyone who has played the dice game craps or the board game *Catan* knows the frustration that this expected value on two dice can generate.

Now you can see where the terms bra and ket come from — they "bracket" an operator to give you expectation values. In fact, the expectation value is such a common thing to find that you often find $\langle\psi|R|\psi\rangle$ abbreviated as $\langle R\rangle$, so for this example,

$$\langle R\rangle = 7$$

Looking at linear operators

An operator A is said to be *linear* if it meets the following condition:

$$A(c_1|\psi\rangle + c_2|\phi\rangle) = c_1 A|\psi\rangle + c_2 A|\phi\rangle$$

What does this equation actually mean? If an operator A acts on the linear sum of two kets (the left side of the equation), then it is equivalent to a linear sum of those same kets being acted on individually by A (the right side of the equation). In other words, a linear operator doesn't do something weird that intertwines the two kets in a complicated way. It allows you to break the two pieces apart and look at them individually, or alternately to combine them together.

I propose that the expression $|\phi\rangle\langle\chi|$ is actually a linear operator. And you can follow these steps — which involve discovering a little more about what happens when you take the products of bras and kets — to prove it:

1. **Take the product of the bra, $\langle\chi|$, and the ket, $c|\psi\rangle$, where c is a complex number.**

   Your result is the following equation:

   $$\langle\chi|c|\psi\rangle = c\langle\chi|\psi\rangle$$

2. **Take the product of the bra, $\langle\chi|$, with the sum of two kets, $|\psi_1\rangle + |\psi_2\rangle$.**

   Your result is the following equation:

   $$\langle\chi|(|\psi_1\rangle + |\psi_2\rangle) = \langle\chi|\psi_1\rangle + \langle\chi|\psi_2\rangle$$

3. **To test that $|\phi\rangle\langle\chi|$ is a linear operator, apply it to a linear combination of bras, where c_1 and c_2 are complex numbers, like so:**

   $$|\phi\rangle\langle\chi|(c_1|\psi_1\rangle + c_2|\psi_2\rangle) = |\phi\rangle\langle\chi|c_1|\psi_1\rangle + |\phi\rangle\langle\chi|c_2|\psi_2\rangle$$

4. **Using the result equation from Step 1, $\langle\chi|c|\psi\rangle = c\langle\chi|\psi\rangle$, you can finally write the equation as**

   $$|\phi\rangle\langle\chi|(c_1|\psi_1\rangle + c_2|\psi_2\rangle) = c_1|\phi\rangle\langle\chi|\psi_1\rangle + c_2|\phi\rangle\langle\chi|\psi_2\rangle$$

Now, if you look at the condition that defines a linear operator, A, at the beginning of this section, you may not immediately recognize that your final equation matches up with it. But if you replace that A in the earlier equation with $|\phi\rangle\langle\chi|$, sure enough, you get exactly this most recent equation!

Going Hermitian with Hermitian operators and adjoints

The *Hermitian adjoint* — also called the adjoint or Hermitian conjugate — of an operator A is denoted $A^\dagger$. To find the Hermitian adjoint, follow these steps:

1. Replace complex constants with their complex conjugates.

The *Hermitian adjoint* of a complex number is the complex conjugate of that number, which is denoted as

$$a^\dagger = a^*$$

2. Replace kets with their corresponding bras, and replace bras with their corresponding kets.

WARNING

You have to exchange the bras and kets when finding the Hermitian adjoint of an operator. As opposed to finding the Hermitian adjoint of a complex number, finding the Hermitian adjoint of an *operator* is not just the same as mathematically finding its complex conjugate.

3. Replace operators with their Hermitian operators.

In quantum mechanics, operators that are equal to their Hermitian adjoints are called Hermitian operators. In other words, an operator is Hermitian if

$$A^\dagger = A$$

Hermitian operators appear throughout the book, and they have special properties. For example, the matrix that represents them may be diagonalized — that is, written so that the only nonzero elements appear along the matrix's diagonal. Also, the expectation value of a Hermitian operator is guaranteed to be a real number, not complex (see the earlier section "I expected that: Finding expectation values").

4. Write your final equation.

$$\left(\langle\psi|A|\phi\rangle\right)^* = \langle\phi|A^\dagger|\psi\rangle$$

REMEMBER

Here are some useful relationships concerning Hermitian adjoints:

>> $(a\mathrm{A})^{\dagger} = a*\mathrm{A}^{\dagger}$

>> $(\mathrm{A}^{\dagger})^{\dagger} = \mathrm{A}$

>> $(\mathrm{A}+\mathrm{B})^{\dagger} = \mathrm{A}^{\dagger} + \mathrm{B}^{\dagger}$

>> $(\mathrm{AB})^{\dagger} = \mathrm{B}^{\dagger}\mathrm{A}^{\dagger}$

>> $(\mathrm{AB}|\psi\rangle)^{\dagger} = \langle\psi|\mathrm{B}^{\dagger}\mathrm{A}^{\dagger}$

Forward and Backward: Finding the Commutator

REMEMBER

The measure of how different the results are when you apply operator A and then B, versus B and then A, is called the operators' *commutator*. In quantum mechanics, the commutator indicates whether you can measure two observable traits at the same time, or whether the two traits have some sort of dependence upon each other in terms of measurement. Here's how you define the commutator of operators A and B:

$$[\mathrm{A, B}] = \mathrm{AB} - \mathrm{BA}$$

Commuting

Two operators commute with each other if their commutator is equal to zero. And when it is, the two observable traits represented by the operators can be measured at the same time. That is, the operators commute when it doesn't make any difference in what order you apply them:

$$[\mathrm{A, B}] = 0$$

Note in particular that any operator commutes with itself:

$$[\mathrm{A, A}] = 0$$

And when operators commute, it's easy to show that the commutator of A, B is the negative of the commutator of B, A:

$$[\mathrm{A, B}] = -[\mathrm{B, A}]$$

It's also true that commutators are linear:

$$[A, B+C+D+\ldots]=[A, B]+[A, C]+[A, D]+\ldots$$

And the Hermitian adjoint of a commutator works this way:

$$[A, B]^{\dagger}=\left[B^{\dagger}, A^{\dagger}\right]$$

You can also find the *anticommutator* — denoted as {A, B} — which is the sum, rather than the difference, of the applied operators:

$$\{A, B\} = AB + BA$$

Finding anti-Hermitian operators

What can you say about the Hermitian adjoint of the commutator of two Hermitian operators? Here's how to find out. First, write the adjoint. Then apply the definition of commutators (see the preceding section). And because you find (in the section "Going Hermitian with Hermitian operators and adjoints," earlier in the chapter) that $(AB)^{\dagger}=B^{\dagger}A^{\dagger}$ and $A=A^{\dagger}$ for Hermitian operators, you can work through the following sequence:

$$[A,B]^{\dagger}=(AB-BA)^{\dagger}=B^{\dagger}A^{\dagger}-A^{\dagger}B^{\dagger}$$

But BA – AB is just –[A, B], so you have the equation,

$$[A, B]^{\dagger}=-[A, B]$$

REMEMBER

A and B here are Hermitian operators. When you take the Hermitian adjoint of an expression and get the same thing back with a negative sign in front of it, the expression is called anti-Hermitian, so the commutator of two Hermitian operators is anti-Hermitian. (And by the way, the expectation value of an anti-Hermitian operator is guaranteed to be completely imaginary.)

Starting from Scratch and Ending Up with Heisenberg

If you read the last few sections, you're now armed with all these new tools: Hermitian operators and commutators. How can you put this knowledge to work? You can come up with the Heisenberg uncertainty principle starting virtually from scratch. (I explain why this uncertainty principle is important in the section "Understanding some relationships by using kets," earlier in the chapter.)

Here's a calculation that takes you from a few basic definitions to the Heisenberg uncertainty principle. This kind of calculation shows how much easier it is to use the basis-less bra and ket notation than the full matrix version of state vectors. (Heisenberg derived his uncertainty principle in 1927, but Dirac notation wouldn't be created until 1939, so the principle was originally derived using the far more cumbersome matrix approach, which Heisenberg also invented.)

You may know how to calculate the uncertainty in a measurement from previous statistics work. You do so by subtracting the squared expectation value from the expectation value of the quantity squared. For Hermitian operators A and B, that gives you

$$\Delta A = \left(\langle A^2 \rangle - \langle A \rangle^2 \right)^{1/2}$$
$$\Delta B = \left(\langle B^2 \rangle - \langle B \rangle^2 \right)^{1/2}$$

Now consider the operators ΔA and ΔB (not the uncertainties), and assume that applying these operators gives you measurement values like this:

$$\Delta A = A - \langle A \rangle$$
$$\Delta B = B - \langle B \rangle$$

Like any operator, they can result in new kets:

$$\Delta A |\psi\rangle = |\chi\rangle$$
$$\Delta B |\psi\rangle = |\phi\rangle$$

Here's the key: The Schwarz inequality (from the earlier section "Understanding some relationships by using kets") gives you

$$\langle \chi | \chi \rangle \langle \phi | \phi \rangle \geq |\langle \chi | \phi \rangle|^2$$

TIP

You can see that the inequality sign, $\geq$, which plays a big part in the Heisenberg uncertainty principle, has already crept into the calculation.

Because $\Delta A^\dagger = \Delta A$ (the definition of a Hermitian operator), you can see that

$$\langle \chi | \chi \rangle = \langle \psi | \Delta A^\dagger \Delta A | \psi \rangle = \langle \psi | \Delta A^2 | \psi \rangle$$

That is, $\langle \chi | \chi \rangle = \langle \Delta A^2 \rangle$. By similar argument reasoning, $\langle \phi | \phi \rangle = \langle \Delta B^2 \rangle$. So, you can rewrite the Schwarz inequality like this:

$$\langle \Delta A^2 \rangle \langle \Delta B^2 \rangle \geq |\langle \Delta A \Delta B \rangle|^2$$

Now, by applying some clever math, and the facts known about Hermitians, you can eventually get the equation into this form:

$$\langle \Delta A^2 \rangle \langle \Delta B^2 \rangle \geq \frac{1}{4} |\langle [A, B] \rangle|^2$$

This combined equation can get you to the Heisenberg uncertainty relation (as introduced in Chapter 3), which looks like this:

$$\Delta x \Delta p_x \geq \frac{\hbar}{2}$$

$$\langle \Delta A^2 \rangle = \Delta A^2$$

Now you can jump back to the earlier combined equation — the one that you want to reflect the Heisenberg uncertainty relation — and apply the relationship you calculated using the expectation value (to both A and B). Substituting this in, and then taking the square root, you're able to get

$$\langle \Delta A^2 \rangle \langle \Delta B^2 \rangle \geq \frac{1}{4} |\langle [A, B] \rangle|^2$$

$$\Delta A^2 \Delta B^2 \geq \frac{1}{4} |\langle [A, B] \rangle|^2$$

$$\Delta A \Delta B \geq \frac{1}{2} \langle [A, B] \rangle$$

Well, well, well. So, the product of two uncertainties is greater than or equal to one-half the absolute value of the commutator of their respective operators? Wow. Is that the Heisenberg uncertainty relation? Well, take a look. In quantum mechanics, the momentum operator (mentioned way earlier in the section "Grooving with Operators") looks like this:

$$P = -i\hbar \nabla$$

And the operator for the momentum in the x direction is

$$P_x = -i\hbar \frac{\partial}{\partial x}$$

So, what's the commutator of the X operator (which just returns the x position of a particle) and P_x? You know that $[X, P_x] = i\hbar$, so from $\Delta A \Delta B \geq \frac{1}{2} |\langle [A, B] \rangle|$, you get this next equation (remember, these are the uncertainties in x and p_x, not the operators):

$$\Delta x \Delta p_x \geq \frac{\hbar}{2}$$

Hot dog! That is the Heisenberg uncertainty relation. I haven't actually constrained the physical world with pure mathematics . . . but I have shown, from a few basic assumptions, that you can't *measure* the physical world with perfect accuracy.

Eigenvectors and Eigenvalues: They're Naturally Eigentastic!

In the previous section, you discover that applying an operator to a ket can result in a new ket. For example, when you apply the operator A, you can get the following result:

$$A|\psi\rangle = |\chi\rangle$$

REMEMBER

To make working with vectors easier, you can work with eigenvectors and eigenvalues (*eigen* is German for "innate" or "natural"). Eigenvectors and eigenvalues are a central concept in linear algebra and allow you to apply operators without changing the state vector. They also help determine the solutions to differential equations. A vector $|\psi\rangle$ is an eigenvector of the operator A if, for some complex constant a,

$$A|\psi\rangle = a|\psi\rangle$$

Note what's happening here: Applying A to one of its eigenvectors gives you that same vector back, multiplied by a number, a. That number is called the eigenvalue of that eigenvector. Although a can be a complex constant, the eigenvalues of Hermitian operators are real numbers, and their eigenvectors are orthogonal (that is, $\langle\psi|\phi\rangle = 0$). For more information, see the section "Understanding some relationships by using kets," earlier in the chapter.

Casting a problem in terms of eigenvectors and eigenvalues can make life a lot easier because applying the operator to its eigenvectors merely gives you the same eigenvector back again, multiplied by its eigenvalue — there's no pesky change of state, so you don't have to deal with a different state vector.

Take a look at this idea, using the R operator from rolling the dice, which is expressed this way in matrix form (see the earlier section "I expected that: Finding expectation values" for more on this matrix):

$$R = \begin{bmatrix} 2 & 0 & 0 & 0 & 0 & 0 & 0 & 0 & 0 & 0 & 0 \\ 0 & 3 & 0 & 0 & 0 & 0 & 0 & 0 & 0 & 0 & 0 \\ 0 & 0 & 4 & 0 & 0 & 0 & 0 & 0 & 0 & 0 & 0 \\ 0 & 0 & 0 & 5 & 0 & 0 & 0 & 0 & 0 & 0 & 0 \\ 0 & 0 & 0 & 0 & 6 & 0 & 0 & 0 & 0 & 0 & 0 \\ 0 & 0 & 0 & 0 & 0 & 7 & 0 & 0 & 0 & 0 & 0 \\ 0 & 0 & 0 & 0 & 0 & 0 & 8 & 0 & 0 & 0 & 0 \\ 0 & 0 & 0 & 0 & 0 & 0 & 0 & 9 & 0 & 0 & 0 \\ 0 & 0 & 0 & 0 & 0 & 0 & 0 & 0 & 10 & 0 & 0 \\ 0 & 0 & 0 & 0 & 0 & 0 & 0 & 0 & 0 & 11 & 0 \\ 0 & 0 & 0 & 0 & 0 & 0 & 0 & 0 & 0 & 0 & 12 \end{bmatrix}$$

The R operator works in 11-dimensional space and is Hermitian, so the operator will have 11 orthogonal eigenvectors and 11 corresponding eigenvalues.

Because R is a diagonal matrix, finding the eigenvectors is easy. You can take unit vectors in the 11 different directions as the eigenvectors. Here's what some of the eigenvectors — the first (ξ_1), second (ξ_2), and then on to the final, eleventh eigenvector (ξ_{11}) — would look like:

$$\xi_1 = \begin{bmatrix} 1 \\ 0 \\ 0 \\ 0 \\ 0 \\ 0 \\ 0 \\ 0 \\ 0 \\ 0 \\ 0 \end{bmatrix} \quad \xi_2 = \begin{bmatrix} 0 \\ 1 \\ 0 \\ 0 \\ 0 \\ 0 \\ 0 \\ 0 \\ 0 \\ 0 \\ 0 \end{bmatrix} \quad \xi_{11} = \begin{bmatrix} 0 \\ 0 \\ 0 \\ 0 \\ 0 \\ 0 \\ 0 \\ 0 \\ 0 \\ 0 \\ 1 \end{bmatrix}$$

Each eigenvector has zero values, except in one location. This means that you can't write any of these vectors as a linear combination of the other vectors. They're linearly independent of each other and, in this case, also orthogonal. They define the basis of this 11-dimensional dice space.

And the eigenvalues? They're the numbers you get when you apply the R operator to an eigenvector. Because the eigenvectors are just unit vectors in all 11 dimensions, the eigenvalues are the numbers on the diagonal of the R matrix: 2, 3, 4, and so on, up to 12.

Understanding how they work

The eigenvectors of a Hermitian operator define a complete set of orthonormal vectors — that is, a complete basis for the state space. (See the section "Understanding some relationships by using kets," earlier in this chapter, for info about orthonormal vectors.) When viewed in this "eigenbasis," which is built of the eigenvectors, the operator in matrix format is diagonal and the elements along the diagonal of the matrix are the eigenvalues.

You might start out with an initial matrix that's quite a mess. Bear in mind that the elements in an operator can also be functions, not just numbers. So far in this chapter, the examples are pretty straightforward because they have only numerical elements in the matrix, but even with numbers, the matrix can get ugly . . . or, as math teachers like to say, "inelegant."

By switching one of these ugly matrices to the basis of eigenvectors for the operator, you diagonalize the matrix into something more like what you've seen, which is much easier to work with. If you want to go into more depth on the mathematics of calculating eigenvalues or eigenvectors, you can also look into *Linear Algebra For Dummies*.

If two or more of the eigenvalues are the same, that eigenvalue is said to be degenerate. So, for example, if three eigenvalues are equal to 6, then the eigenvalue 6 is threefold degenerate.

Here's another cool thing: If two Hermitian operators, A and B, commute, and if A doesn't have any degenerate eigenvalues, then each eigenvector of A is also an eigenvector of B. (See the earlier section "Forward and Backward: Finding the Commutator" for more on commuting.)

Finding eigenvectors and eigenvalues

So, given an operator in matrix form, how do you find its eigenvectors and eigenvalues? This is the equation you want to solve:

$$A|\psi\rangle = a|\psi\rangle$$

And you can rewrite this equation as the following:

$$(A - aI)|\psi\rangle = 0$$

"I" represents the *identity matrix*, with 1s along its diagonal and 0s otherwise:

$$I = \begin{bmatrix} 1 & 0 & 0 & 0 & 0 & 0 & 0 & 0 & 0 & 0 & \cdots \\ 0 & 1 & 0 & 0 & 0 & 0 & 0 & 0 & 0 & 0 & \cdots \\ 0 & 0 & 1 & 0 & 0 & 0 & 0 & 0 & 0 & 0 & \cdots \\ 0 & 0 & 0 & 1 & 0 & 0 & 0 & 0 & 0 & 0 & \cdots \\ 0 & 0 & 0 & 0 & 1 & 0 & 0 & 0 & 0 & 0 & \cdots \\ 0 & 0 & 0 & 0 & 0 & 1 & 0 & 0 & 0 & 0 & \cdots \\ & & & & \vdots & & & & & \cdots \end{bmatrix}$$

REMEMBER

The solution to $(A - aI)|\psi\rangle = 0$ exists only if the *determinant* (a scalar value you calculate from the elements of a matrix) of the matrix $A - aI$ is 0:

$$\det(A - aI) = 0$$

Finding eigenvalues

Any values of a that satisfy the equation $\det(A - aI) = 0$ are eigenvalues of the original equation. Try to find the eigenvalues and eigenvectors of the following matrix:

$$A = \begin{bmatrix} -1 & -1 \\ 2 & -4 \end{bmatrix}$$

First, convert the matrix into the form $A - aI$. which has only 1s in the diagonal positions and 0s everywhere else. Since you are subtracting aI, you will be subtracting a from the diagonal matrix elements of A:

$$A - aI = \begin{bmatrix} -1-a & -1 \\ 2 & -4-a \end{bmatrix}$$

Next, find the determinant:

$$\det(A - aI) = (-1-a)(-4-a) + 2$$
$$\det(A - aI) = a^2 + 5a + 6$$

And you can factor this equation as follows:

$$\det(A - aI) = a^2 + 5a + 6 = (a+2)(a+3)$$

You know that $\det(A - aI) = 0$, so the eigenvalues of A are the roots of this equation; namely, $a_1 = -2$ and $a_2 = -3$.

Finding eigenvectors

How about finding the eigenvectors? To find the eigenvector corresponding to a_1 (see the preceding section), substitute a_1 — the first eigenvalue, -2 — into the matrix in the form $A - aI$:

$$A - aI = \begin{bmatrix} -1-a & -1 \\ 2 & -4-a \end{bmatrix} = \begin{bmatrix} 1 & -1 \\ 2 & -2 \end{bmatrix}$$

So you have the value of $A - aI$ that you can use:

$$(A - aI)|\psi\rangle = 0 \implies \begin{bmatrix} 1 & -1 \\ 2 & -2 \end{bmatrix} \begin{bmatrix} \psi_1 \\ \psi_2 \end{bmatrix} = \begin{bmatrix} 0 \\ 0 \end{bmatrix}$$

Because every row of this matrix equation must be true, you know that $\psi_1 = \psi_2$. Both elements of the eigenvector will be the same. The easiest values to use are 1 for each element (you can always multiply it by a value later when using it as a basis), giving you the following eigenvector for the first eigenvalue:

$$\begin{bmatrix} 1 \\ 1 \end{bmatrix}$$

How about the eigenvector corresponding to a_2? Plugging a_2, -3, into the matrix in $A - aI$ form, you get the following:

$$A - aI = \begin{bmatrix} 2 & -1 \\ 2 & -1 \end{bmatrix}$$

$$(A - aI)|\psi\rangle = 0 \implies \begin{bmatrix} 2 & -1 \\ 2 & -1 \end{bmatrix} \begin{bmatrix} \psi_1 \\ \psi_2 \end{bmatrix} = \begin{bmatrix} 0 \\ 0 \end{bmatrix}$$

This gives the equations

$$2\psi_1 - \psi_2 = 0$$
$$\psi_1 = \psi_2 \div 2$$

You can solve these to show that the second element will always be twice the first, meaning that the eigenvector corresponding to a_2 is

$$\begin{bmatrix} 1 \\ 2 \end{bmatrix}$$

Preparing for the Inversion: Simplifying with Unitary Operators

Applying the *inverse* of an operator undoes the work the operator did (and that's why it's called an inverse), so when an inverse operates on itself, it becomes the identity matrix:

$$A^{-1}A = AA^{-1} = I$$

Sometimes, finding the inverse of an operator is helpful, such as when you want to solve equations like $Ax = y$. Solving for x is easy if you can find the inverse of A: $x = A^{-1}y$.

However, finding the inverse of a large matrix often isn't easy, so quantum physics calculations are sometimes limited to working with unitary operators, U, where the operator's inverse is equal to its adjoint, $U^{-1} = U^{\dagger}$. (To find the adjoint of an operator, A, you find the transpose by interchanging the rows and columns, A^{T}. Then take the complex conjugate, $A^{T*} = A^{\dagger}$.) This gives you the following equation:

$$U^{\dagger}U = UU^{\dagger} = 1$$

The product of two unitary operators, U and V, is also unitary because

$$(UV)(UV)^{\dagger} = (UV)(V^{\dagger}U^{\dagger}) = U(VV^{\dagger})U^{\dagger} = UU^{\dagger} = I$$

When you use unitary operators, kets and bras transform this way:

>> $|\psi'\rangle = U|\psi\rangle$

>> $\langle\psi'| = \langle\psi|U^{\dagger}$

And you can transform other operators using unitary operators like this:

$$A' = UAU^{\dagger}$$

Note that the preceding equations also mean the following:

>> $|\psi\rangle = U^{\dagger}|\psi'\rangle$

>> $\langle\psi| = \langle\psi'|U$

>> $A = U^{\dagger}A'U$

REMEMBER

Here are some properties of unitary transformations:

>> If an operator is Hermitian, then its unitary transformed version, $A' = UAU^\dagger$, is also Hermitian.

>> The eigenvalues of A and its unitary transformed version, $A' = UAU^\dagger$, are the same.

>> Commutators that are equal to complex numbers are unchanged by unitary transformations: $[A', B'] = [A, B]$.

Comparing Matrix and Continuous Representations

Werner Heisenberg developed the matrix-oriented view of quantum physics that you've been using so far in this chapter. It's sometimes called matrix mechanics. The matrix representation is fine for many problems, but sometimes you have to go past it, as you discover in this section.

REMEMBER

The Heisenberg matrix technique is best used for physical systems with well-defined energy states, such as harmonic oscillators. The Schrödinger way of looking at things, wave mechanics, uses wave functions, mostly in the position basis, to reduce questions in quantum physics to a differential equation.

One of the central goals of quantum mechanics problems is to calculate the energy levels of a system. The energy operator is called the Hamiltonian, H, and finding the energy levels of a system breaks down to finding the eigenvalues related to the situation:

$$H|\psi\rangle = E|\psi\rangle$$

Here, E is an eigenvalue of the H operator. This same equation can be written as a matrix:

$$\begin{bmatrix} H_{11}-E & H_{12} & H_{13} & H_{14} & \cdots \\ H_{21} & H_{22}-E & H_{23} & H_{24} & \cdots \\ H_{31} & H_{32} & H_{33}-E & H_{34} & \cdots \\ H_{41} & H_{42} & H_{43} & H_{44}-E & \cdots \\ & & \vdots & & \end{bmatrix} \begin{bmatrix} \psi_1 \\ \psi_2 \\ \psi_3 \\ \psi_4 \\ \vdots \end{bmatrix} = 0$$

The allowable energy levels of the physical system are the eigenvalues E, which satisfy this equation. If you have a discrete basis of eigenvectors — if the number of energy states is finite — then you can set the determinant equal to zero to find those eigenvalues.

But what if the number of energy states is infinite? In that case, you can no longer use a discrete basis for your operators and bras and kets — you must use a continuous basis.

Going continuous with calculus

Representing quantum mechanics in a continuous basis is an invention of the physicist Erwin Schrödinger. In the continuous basis, summations become integrals. For example, take the following relation, where I is the identity matrix:

$$\sum_{n=1}^{\infty} |\phi_n\rangle\langle\phi_n| = I$$

It becomes the following as an integral:

$$\int d\phi |\phi\rangle\langle\phi| = I$$

And every ket $|\psi\rangle$ can be expanded in a basis of other kets, $|\phi_n\rangle$, like this:

$$|\psi\rangle = \int d\phi |\phi\rangle\langle\phi|\psi\rangle$$

Doing the wave

Take a look at the position operator, R, in a continuous basis. Applying this operator gives you r, the position vector:

$$R|\psi\rangle = r|\psi\rangle$$

In this equation, applying the position operator to a state vector returns the locations, r, that a particle may be found at. You can expand any ket in the position basis like this:

$$|\psi\rangle = \int d^3r |r\rangle\langle r|\psi\rangle$$

And this becomes

$$|\psi\rangle = \int d^3r \psi(r)|r\rangle$$

REMEMBER

Here's a very important thing to understand: $\psi(r) = \langle r|\psi\rangle$ is the wave function for the state vector $|\psi\rangle$ — it's the ket's representation in the position basis. Or in common terms, it's just a function where the quantity $|\psi(r)|^2 d^3r$ represents the probability that the particle will be found in the region d^3r at r.

The wave function is the foundation of what's called wave mechanics, as opposed to matrix mechanics. What's important to realize is that when you talk about representing physical systems in wave mechanics, you don't use the basis-less bras and kets of matrix mechanics; rather, you usually use the wave function — that is, bras and kets in the position basis.

Therefore, you go from talking about $|\psi\rangle$ to $\langle r|\psi\rangle$, which equals $\psi(r)$. This wave function is the central mathematical object of interest in quantum physics, and it's just a ket in the position basis. So, in wave mechanics, $H|\psi\rangle = E|\psi\rangle$ becomes the following:

$$\langle r|H|\psi\rangle = E\langle r|\psi\rangle$$

You can write this as the following:

$$\langle r|H|\psi\rangle = E\langle r|\psi\rangle$$

But what is $\langle r|H|\psi\rangle$? It's equal to $H\psi(r)$. The Hamiltonian operator, H, is the total energy of the system, kinetic ($p^2/2m$) plus potential ($V(r)$), so you get the following equation:

$$H = \frac{p^2}{2m} + V(r)$$

But the momentum operator is

$$P|\psi\rangle = -i\hbar \frac{\partial}{\partial x}|\psi\rangle \boldsymbol{i} - i\hbar \frac{\partial}{\partial y}|\psi\rangle \boldsymbol{j} - i\hbar \frac{\partial}{\partial z}|\psi\rangle \boldsymbol{k}$$

Therefore, substituting the momentum operator for p gives you this:

$$H = \frac{-\hbar^2}{2m}\left(\frac{\partial^2}{\partial x^2} + \frac{\partial^2}{\partial y^2} + \frac{\partial^2}{\partial z^2}\right) + V(r)$$

Using the Laplacian operator, you get this equation:

$$\nabla^2|\psi\rangle = \frac{\partial^2}{\partial x^2}|\psi\rangle + \frac{\partial^2}{\partial y^2}|\psi\rangle + \frac{\partial^2}{\partial z^2}|\psi\rangle$$

You can rewrite this equation as the following (called the *Schrödinger equation*):

$$H\psi(r) = \frac{-\hbar^2}{2m}\nabla^2\psi(r) + V(r)\psi(r) = E\psi(r)$$

So, in the wave mechanics view of quantum physics, you're now working with a differential equation instead of multiple matrices of elements. This all came from working in the position basis, $\psi(r) = \langle r|\psi\rangle$, instead of just $|\psi\rangle$.

The quantum physics in the rest of the book is largely about solving this differential equation for a variety of potentials, V(r). That is, your focus is on finding the wave function that satisfies the Schrödinger equation for various physical systems. When you solve the Schrödinger equation for $\psi(r)$, you can find the allowed energy states for a physical system, as well as the probability that the system will be in a certain position state.

Note that, besides wave functions in the position basis, you can also give a wave function in the momentum basis, $\psi(p)$, or in any number of other bases.

Chapter **8**

Getting Stuck in Energy Wells

I n a classic television series, the dog named Lassie always came to the rescue. There was usually an exchange where the dog was barking, and people would have to try to interpret her barks. "What's the matter, Lassie? Timmy's stuck in the well?"

Fortunately, Timmy isn't stuck in a well today, but instead, you find electrons stuck in wells. *Potential energy wells*, that is. And unfortunately for the electrons (but fortunately for Timmy, I suppose), these sorts of wells bind the electrons in a way that they can't get out of just by grabbing onto a rope lowered down the well.

In this chapter, you see quantum physics at work solving energy-well problems in one dimension. In quantum physics, because energy is quantized, only certain energy states are allowed. You analyze particles trapped in potential wells and solve for the *allowable energy states* using quantum physics. In classical physics, trapped particles aren't restricted to any particular energy spectrum, but when the world gets microscopic, quantum physics takes over.

If you haven't read the previous chapter, then you haven't seen my presentation of how to derive the Schrödinger equation for a wave function, which is the key equation that I put to use in this chapter. And so, I advise that you find this content in Chapter 7. By using the Schrödinger equation, you can solve for the wave function, $\psi(x)$, and the total energy levels, E, for the particles in your energy wells:

$$\frac{-\hbar^2}{2m}\nabla^2\psi(r) + V(r)\psi(r) = E\psi(r)$$

Note: In addition to the Schrödinger equation (found in Chapter 7), this chapter references basic trigonometric functions (such as sine and cosine), second-order differential equations, and integrals from calculus. A familiarity with these mathematical functions and related calculations gives you a good basis for the development of this chapter's equations. I don't go into detail, for example, on how you find solutions to second-order differential equations. That's outside the scope of this book. (Although *Differential Equations For Dummies* does exist, for those who are interested.)

Looking into a Square Well

So what is a potential energy well? Think about the well in terms of gravity. Suppose that you have a grape that you drop into a large bowl, like a mixing bowl with steep sides. The grape will naturally come to rest at the bottom of the bowl. It may move around within the bowl a bit (while you're walking around carrying the bowl), but it won't jump out of the bowl. Even if you jostle the bowl around so that the grape begins rolling, the grape is unlikely to roll out. In order to roll out of the bowl, the grape would need to overcome the gravitational potential energy that's represented by the sides of the bowl. That grape must somehow get a lot of energy to come out of the bowl.

Similarly, when you consider potential wells in this chapter, you consider particles that are trapped between steep "walls" of potential energy. The particle has to move around inside the well, and it can't get out — at least not in the cases covered in this chapter.

To simplify the example, I present a particle in a single dimension (as opposed to the three-dimensional grape-in-a-bowl example), so the particle can only move linearly left or right. And I cheat a bit by focusing initially on a convenient type of energy well. A *square well* is a potential (that is, a potential energy well) that forms a square shape, as shown in Figure 8-1.

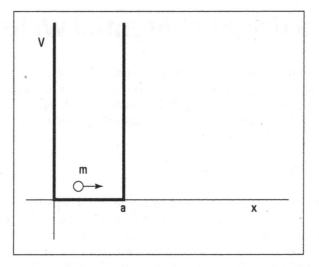

FIGURE 8-1:
A square well.

Another convenience to simplify the example is that I assume the potential energy cannot be overcome. Infinite energy may not be a reality yet, but for these purposes, it's a useful approximation to model a really big energy relative to the energy of the trapped particle. The energy well has some horizontal distance, which I just call a. One neat thing in physics is that you can usually set your experiment in coordinates that make the math most convenient, so I assume that the far-left wall of the energy well is at $x = 0$.

For a square energy well of any size, a, the potential, or $V(x)$, goes to infinity at $x = 0$ and $x > a$ (where x is distance), like this:

>> $V(x) = \infty$, where $x < 0$ or $x > a$

>> $V(x) = 0$, where $0 \leq x \leq a$

Does the particle just sort of roll back and forth on the bottom of the square well, in that region from 0 to a where $V(x) = 0$? Not exactly. Remember that in quantum physics, energy takes on discrete quantized values. The particle is in a bound state, and its wave function depends on its energy. You can find a discussion of the discrete quantized values and wave functions in Chapter 4.

Later in this chapter (in the section "Trapping Particles in Infinite Square Potential Wells"), you find out how to derive the wave function and energy states of particles in a square well. For now, it's useful to just understand the concept of what's going on in potential wells.

Trapping Particles in Potential Wells

When talking about trapping a particle in potential wells, a key feature to consider is how much energy the particle has. Since the particle in question doesn't have a means of grabbing more energy from outside of itself, the particle's total energy, E, is equal to the sum of its kinetic energy and potential energy (V), as shown in this equation:

$$\frac{p^2}{2m} + V = E$$

Take a look at the potential well in Figure 8-2. Notice the dip, or *well*, in the potential path (drawn as a bold, black line), which means that particles can be trapped in the dip if they don't have enough energy to escape. This figure looks a lot more like the grape-in-a-bowl example (from the preceding section), rather than the infinitely steep walls of the square well (refer to Figure 8-1).

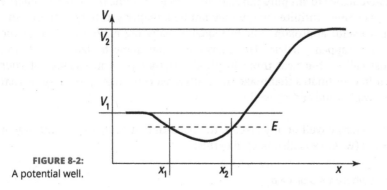

FIGURE 8-2:
A potential well.

In this section, you take a look at the various possible states that a particle with energy E can take in the potential well depicted in Figure 8-2. Quantum-mechanically speaking, those states are of two kinds: bound and unbound. This section offers an overview of these states.

Binding particles in potential wells

Bound states happen when the particle isn't free to travel to infinity. In other words, the particle is confined to the potential well. A particle traveling in the potential well shown in Figure 8-2 is bound if its energy, E, is less than V_1 (which also makes it less than V_2). In that case, the particle moves between x_1 and x_2. A particle trapped in such a well is represented by a wave function, and you can solve the Schrödinger equation for the allowed wave functions and the allowed energy states. The Schrödinger equation is a *second-order differential equation*, which means it requires two boundary conditions to solve the problem completely.

REMEMBER

Bound states are discrete — that is, they form a series of discrete energy levels, as opposed to a continuous spectrum of energies. The Schrödinger equation gives you those states. In addition, in one-dimensional problems, no two energy levels are the same. Or, to use the technical term, the energy levels of a bound state are not *degenerate*.

Escaping from potential wells

If a particle's energy, E, is greater than the potential V_1 (refer to Figure 8-2), the particle can escape from the potential well. There are two possible cases: $V_1 < E < V_2$ and $E > V_2$. This section looks at them separately.

Case 1: Energy between the two potentials ($V_1 < E < V_2$)

If $V_1 < E < V_2$, the particle in the potential well has enough energy to overcome the barrier on the left but not on the right. The particle is free to move to negative infinity, so its allowed x region is between $-\infty$ and x_1.

In this case, the allowed energy values are continuous, not discrete, because the particle isn't completely bound. The energy eigenvalues are not degenerate — that is, no two energy solutions are the same (see Chapter 7 for more on eigenvalues).

The Schrödinger equation is a second-order differential equation, so it has two linearly independent solutions; however, in this case, only one of those solutions doesn't *diverge* (doesn't go to infinity). If a solution goes to infinity, then it can't possibly represent a real physical solution to the problem, because infinities do not exist in actual physical reality. Part of the work being done in quantum physics is applying boundary conditions that help constrain the equation so that it is finite.

The wave equation in this case oscillates for $x < x_2$ and decays rapidly for $x > x_2$.

Case 2: Energy greater than the higher potential ($E > V_2$)

If $E > V_2$, the particle isn't bound at all and is free to travel from negative infinity to positive infinity.

The energy spectrum is continuous, and the wave function turns out to be a sum of one function moving to the right and one moving to the left. The energy levels of the allowed spectrum are therefore *doubly degenerate* (go to infinity in both ways).

That's all the overview you need — time to start solving the Schrödinger equation for various different potentials, starting with the easiest of all: infinite square wells.

Trapping Particles in Infinite Square Potential Wells

Infinite square wells, in which the walls go to infinity, are a favorite in quantum physics problems because they're relatively easy to solve. (The emphasis is on *relatively*.) And in many cases, using an infinite potential is close enough to get a good first approximation for the solution. You can explore the quantum–physics take on these problems in this section.

Finding the wave function equation

Most calculations in quantum physics come down to being able to write and interpret the correct version of the quantum wave function. In this section, I walk you through the steps to determine the wave function for this situation (particles in infinite square wells). Toward the end of the steps, you also find out the allowable energy levels that the resulting equation defines for this situation.

1. **Determine your coordinates to write the Schrödinger equation.**

   Refer to the infinite square well that appears in Figure 8-1. Remember that the square well's potential, V(x), behaves according to the following equalities, based on the position x:

   - V(x) = ∞, where $x < 0$ or $x > a$

   - V(x) = 0, where $0 \leq x \leq a$

   To begin analyzing this situation, you need to break down the Schrödinger equation a bit. Begin with the equation in three dimensions, but in this case, you really just want to look at it in a single dimension, so instead of position r, change it to position x. The Laplacian operator ∇^2 (refer to Chapter 7) behaves like a second-order differential on x, resulting in this transformation:

   $$\frac{-\hbar^2}{2m}\nabla^2\psi(r) + V(r)\psi(r) = E\psi(r)$$

   $$\frac{-\hbar^2}{2m}\frac{d^2}{dx^2}\psi(x) + V(x)\psi(x) = E\psi(x)$$

If you want more details on the different coordinate options, I go into depth on rectangular coordinates in Chapter 12 and spherical coordinates in Chapter 13.

REMEMBER

Notice that E in the Schrödinger equation. Because this energy is in a quantum situation, it can only take on discrete values. The allowable energy states will be all solutions for E that work in this equation. Toward the end of this process, when you've got enough of the wave function, you can calculate those allowed energy levels.

2. **Apply specific constraints and simplify, if possible.**

Now you have the Schrödinger equation with the variables you want (even if you don't know yet why you want it that way). Tackle the easiest parts first.

For a trapped particle, V(x) = 0 inside the well, and the entire term in the middle becomes 0 (because you're multiplying by zero). The equation becomes much simpler:

$$\frac{-\hbar^2}{2m}\frac{d^2}{dx^2}\psi(x) = E\psi(x)$$

3. **Restructure the Schrödinger equation into a solvable form.**

Because the equation still has a lot of constant terms, physicists do often like to collect them together, when possible, to tidy things up.

TIP

In general, you want to get the equation equal to zero. Then you want to see whether you have groups of constants in this new equation that you can combine into a single constant. Knowledge of differential equations is useful here because it tells you which forms of equations you are able to find answers to.

This part of the calculation is just basic algebra (even though you find differentials sitting in the middle of the equation), so you just step through modifications to get your equation to look a little different:

$$\frac{-\hbar^2}{2m}\frac{d^2}{dx^2}\psi(x) = E\psi(x)$$

$$\left(\frac{2m}{-\hbar^2}\right)\frac{-\hbar^2}{2m}\frac{d^2}{dx^2}\psi(x) = \left(\frac{2m}{-\hbar^2}\right)E\psi(x)$$

$$\frac{d^2}{dx^2}\psi(x) = \frac{2mE}{-\hbar^2}\psi(x)$$

$$\frac{d^2}{dx^2}\psi(x) + \frac{2mE}{\hbar^2}\psi(x) = 0$$

$$\frac{d^2}{dx^2}\psi(x) + k^2\psi(x) = 0$$

where $k^2 = \dfrac{2mE}{\hbar^2}$.

This new variable, k, is called the *wave number*. And k is related to the goal of finding E, so when you find a solution that includes k, you can work from that to figure out E.

4. Solve the differential equation for ψ.

So now you have a second-order differential equation to solve for the wave function of a particle trapped in an infinite square well.

$$\frac{d^2}{dx^2}\psi(x) + k^2\psi(x) = 0$$

Solving this differential equation is straightforward. You get two independent solutions:

$$\psi_1(x) = A\sin(kx)$$
$$\psi_2(x) = B\cos(kx)$$

where A and B are constants that are yet to be determined.

REMEMBER

The general solution of $\dfrac{d^2}{dx^2}\psi(x) + k^2\psi(x) = 0$ is the sum of the two independent solutions, $\psi_1(x)$ and $\psi_2(x)$:

$$\psi(x) = A\sin(kx) + B\cos(kx)$$

5. Use the boundary conditions and normalization to find the constants.

What are the boundary conditions? At the boundaries of the infinite square well, the potential energy is infinite, which means that there's no possibility that the particle exists at the boundary itself. You defined the infinite square well so that this is the situation. And the wave function represents the probability that the particle exists at a given point, so if the probability is zero at the boundaries ($x = 0$ and $x = a$), then

- $\psi(0) = 0$

- $\psi(a) = 0$

Using some simple trigonometry and the equations from Step 4 at the boundary values, solve for B. Then use that solution to simplify the second equation.

$$\psi(0) = A\sin(k0) + B\cos(k0) = 0 \Rightarrow B\cos(0) = 0 \Rightarrow B = 0$$
$$\psi(a) = A\sin(ka) + B\cos(ka) = 0 \Rightarrow A\sin(ka) = 0$$

One simple solution would be to consider the case where A = 0. But if both A and B are 0, then the whole wave function equals 0, and the particle doesn't exist anywhere. Since A ≠ 0, consider what other solutions will give a result of 0. I walk you through how to find A in the later section, "Normalizing the wave function."

Before getting there, though, you need to determine the value of k (see the next section) by calculating the energy levels for a particle trapped in a potential well.

Determining the energy levels

The final form of that second equation (from Step 5 of the preceding section) gives you some information about the value of k. From basic trigonometry, you know that the sine is zero when its argument is a multiple of π. In other words, $\sin(ka) = 0$ anytime that:

$$ka = n\pi \quad n = 1, 2, 3 \dots$$

TIP

Note that although $n = 0$ is technically a solution, you've already defined $\psi(0) = 0$ as part of the boundary. It's not a physical solution. The physical solutions begin with $n = 1$.

Now you have found an expression that represents all values of k that fit the quantum situation described. You can rewrite that equation a bit to solve for k:

$$k = \frac{n\pi}{a} \quad n = 1, 2, 3 \dots$$

And because $k^2 = 2mE / \hbar^2$, you have the following equation, where $n = 1, 2, 3, \dots$, as the allowed energy states. These are quantized states corresponding to the quantum numbers 1, 2, 3, and so on:

$$\frac{2mE}{\hbar^2} = \frac{n^2\pi^2}{a^2} \Rightarrow E = \frac{n^2\hbar^2\pi^2}{2ma^2}$$

Congratulations! You just found the formula for the allowable energy states for a particle contained in a square well. You can apply this equation to any specific particle, of mass m, contained in a square potential well of size a.

Note that the first physical state corresponds to $n = 1$, which gives you this next equation, which is the lowest physical state that the particles in an infinite square well can occupy:

$$E = \frac{\hbar^2\pi^2}{2ma^2}$$

Evaluating E for higher values of n would represent higher energy states of the particles.

Just for kicks, put some numbers into this energy equation, assuming that you have an electron, mass 9.11×10^{-31} kilograms, confined to an infinite square well with a width of the order of the *Bohr radius* (the average radius of an electron's orbit in a hydrogen atom); say, $a = 3.00 \times 10^{-10}$ meters. The preceding equation gives you this energy for the ground state:

$$\frac{\left(1.0 \times 10^{-34}\right)^2}{2\left(9.11 \times 10^{-31}\right)} \frac{\left(3.14\right)^2}{\left(3.00 \times 10^{-10}\right)} = 6.69 \times 10^{-18} \text{ Joules}$$

That's a very small amount of energy, about 4.2 *electron volts* (eV — the amount of energy one electron gains falling through 1 volt). Even so, it's already on the order of the energy of an electron in the ground state of a hydrogen atom (13.6 eV), so you can say you're certainly in the right quantum physics ballpark now.

Normalizing the wave function

After you calculate the energy levels for the particle trapped in a square well, you can look a little more specifically at the wave function that describes this situation. What you want to do is *normalize* the wave function, which is to make sure that the probability represented by the wave function matches physical reality.

Using knowledge of infinite boundary conditions, you can find a wave function (see the previous section). You can take that wave function equation and substitute in the value for k (also in the previous section) in physical allowed states.

$$\psi(a) = A \sin(ka) \Rightarrow \psi(a) = A \sin\left(\frac{n\pi x}{a}\right)$$

The wave function is a sine wave, going to zero at $x = 0$ and $x = a$. You can see the first two wave functions plotted in Figure 8-3.

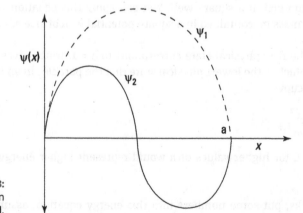

ψ_1

$\psi(x)$

ψ_2

a

x

FIGURE 8-3:
Wave functions in
a square well.

You still don't know the value of A, though, and what you want to do next is to find this value. In this situation, assume that the particle is somewhere within the well. It should be, after all; you put it there, and it doesn't have any way to get out! (If you're about to ask about particles escaping, hold your horses! I explain quantum tunneling later in the chapter. For now, just assume that the particle is somewhere in this well.) When the particle is somewhere in the well and the wave function represents the probability of it being at any point, the sum of that probability across the well should equal 1 (see Chapters 2, 3, and 7 for more information about probabilities).

This process is called *normalizing the wave function*. In a normalized function, the probability of finding the particle between x and dx, |ψ(x)|²dx, adds up to 1 when you integrate over the entire square well, x = 0 to x = a:

$$1 = \int_0^a |\psi(a)|^2 \, dx$$

Substituting for ψ(x) gives you the following, which you can integrate, solve for A, and then substitute that A value back into the wave function equation.

$$1 = |A|^2 \int_0^a \sin^2\left(\frac{n\pi x}{a}\right) dx$$

$$1 = |A|^2 \left(\frac{a}{2}\right)$$

$$A = \left(\frac{2}{a}\right)^{\frac{1}{2}}$$

$$\psi(x) = \sqrt{\frac{2}{a}} \sin\left(\frac{n\pi x}{a}\right) \quad n = 1, 2, 3...$$

And voila . . . that's the normalized wave function for a particle in an infinite square well!

Adding time dependence to wave functions

So far, the calculations in this chapter ignore the element of time, and that's fine, but there is a time-dependent Schrödinger equation that's worth a look because in the real universe (sometimes), things change over time. The big change of note, mathematically speaking, is that the wave function is no longer only a function of position r (or x, in the one-dimensional cases in this book), but is also dependent upon time t. The time-dependent equation looks like this:

$$i\hbar \frac{\partial}{\partial t} \psi(r, t) = H\psi(r, t)$$

The H is the Hermitian Hamiltonian operator, which you can look up in Chapter 7. Applying it, and then focusing on only one dimension, yields another form of the time-dependent Schrödinger equation:

$$i\hbar\frac{\partial}{\partial t}\psi\left(x,\,t\right)=\frac{-\hbar^2}{2m}\frac{\partial^2}{\partial x^2}\psi\left(x,\,t\right)+V\left(x,\,t\right)\psi\left(x,\,t\right)$$

This equation is simpler than it looks, however, because the potential energy doesn't change with time. In fact, because E is constant and the Hermitian Hamiltonian operator just returns the constant energy, the original time-dependent Schrödinger equation becomes

$$i\hbar\frac{\partial}{\partial t}\psi\left(x,\,t\right)=E\psi\left(x,\,t\right)$$

The resulting equation makes life a lot simpler — it's easy to solve the time-dependent Schrödinger equation if you're dealing with a constant potential energy. Just sprinkle in some differential equation magic (and maybe look it up in a table in a differential equation textbook). In this case, the solution is

$$\psi\left(x,\,t\right)=\psi\left(x\right)e^{-iEt/\hbar}$$

REMEMBER

Neat. When the potential doesn't vary with time, the solution to the time-dependent Schrödinger equation simply becomes $\psi(x)$, the spatial part (dependent only on x, but not on t), multiplied by $e^{-iEt/\hbar}$, the time-dependent part (dependent only on t, but not on x).

So, when you add in the time-dependent part to the time-independent wave function (see the section "Normalizing the wave function," earlier in the chapter), you get the time-dependent wave function, which looks like this:

$$\psi\left(x,t\right)=\sqrt{\frac{2}{a}}\sin\left(\frac{n\pi x}{a}\right)e^{-iEt/\hbar}\qquad n=1,\,2,\,3\dots$$

The energy of the nth quantum state (defined in the section "Determining the energy levels," earlier in the chapter) is

$$E=\frac{n^2\hbar^2\pi^2}{2ma^2}\qquad n=1,\,2,\,3\dots$$

Therefore, the resulting time-dependent wave function is

$$\psi\left(x,t\right)=\sqrt{\frac{2}{a}}\sin\left(\frac{n\pi x}{a}\right)\exp\left(-\frac{in^2\hbar\pi^2t}{2ma^2}\right)\qquad n=1,\,2,\,3\dots$$

TIP

The term *exp* in the preceding equation represents the natural exponential function, *e*, but it's formatted this way so that you can more easily read everything in the brackets, which would be difficult if they were in an exponent.

Shifting to symmetric square well potentials

The standard infinite square well extends from $x = 0$ to $x = a$. But what if you want to shift the coordinates so that the square well is symmetric around the origin instead? That is, the new infinite square well looks like this:

>> $V(x) = \infty$, where $x < -\frac{a}{2}$ or $x > \frac{a}{2}$

>> $V(x) = 0$, where $-\frac{a}{2} \geq x \geq \frac{a}{2}$

You can translate from this new square well to the old one by adding $a/2$ to x, which means that you can write the wave function for the new square well in this equation as follows:

$$\psi(x) = \sqrt{\frac{2}{a}} \sin\left(\frac{n\pi}{a}\left(x + \frac{a}{2}\right)\right) \quad n = 1, 2, 3...$$

Doing a little trig gives you the following equations:

$$\psi(x) = \sqrt{\frac{2}{a}} \cos\frac{n\pi x}{a} \quad n = 1, 3, 5...$$

$$\psi(x) = \sqrt{\frac{2}{a}} \sin\frac{n\pi x}{a} \quad n = 2, 4, 6...$$

The result is a mix of sines and cosines. I leave it to you to figure out the bound states and to determine whether any patterns exist among them. (*Hint:* Compare the symmetry of the cosine versus sine states.)

Limited Potential: Taking a Look at Particles and Potential Steps

Truly infinite potentials are hard to come by. The calculations in previous sections are the simplest starting points, and they often provide good approximations for real-world situations.

In this section, you look at some real-world examples, where the potential is set to some finite V_0, not infinity. For example, in Figure 8-4 a particle is traveling toward a point where there's a sudden increase in potential energy, called a *potential step*. Currently, the particle is in a region where $V = 0$, but it'll soon be in the region where $V = V_0$.

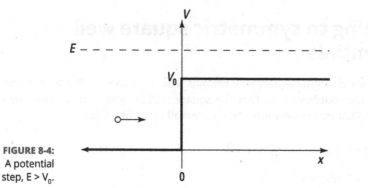

FIGURE 8-4:
A potential
step, E > V_0.

You have two cases to look at here in terms of E, the energy of the particle:

» **E > V_0:** Classically, when E > V_0, you expect the particle to be able to continue on to the region $x > 0$.

» **E ≤ V_0:** When E ≤ V_0, you'd expect the particle to bounce back and not be able to get to the region $x > 0$ at all.

Assuming the particle has plenty of energy

Start with the case where the particle's energy, E, is greater than the potential V_0. From a quantum physics point of view, find the Schrödinger equation by following the first three steps of an earlier section, "Finding the wave function equation":

1. **Determine your coordinates to write the Schrödinger equation.**

2. **Apply specific constraints; simplify if possible.**

3. **Restructure the Schrödinger equation into a solvable form.**
   - For the region $x < 0$: $\dfrac{d^2\psi_1}{dx^2}(x) + k_1^2\psi_1(x) = 0$ and $k_1^2 = \dfrac{2mE}{\hbar^2}$
   - For the region $x > 0$: $\dfrac{d^2\psi_2}{dx^2}(x) + k_2^2\psi_2(x) = 0$ and $k_2^2 = \dfrac{2m(E-V_0)}{\hbar^2}$

   In other words, k varies by region, as illustrated in Figure 8-5. This difference is based upon the size of the potential step, V_0.

4. **Solve the differential equations for ψ.**

   The two equations in Step 3 are second-order differential equations, exactly like those in a section earlier in this chapter, "Finding the wave function equation." As shown in Step 4 of that section, the most general solution is the following:

   $\psi_1(x) = Ae^{ik_1x} + Be^{-ik_1x}$, where $x < 0$

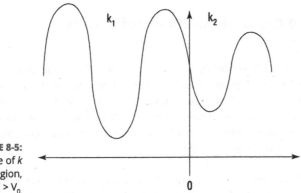

FIGURE 8-5:
The value of k
by region,
where $E > V_0$.

And for the region $x > 0$, you get the same solution, but use different coefficients since it's a different equation:

$\psi_2(x) = Ce^{ik_2x} + De^{-ik_2x}$, where $x > 0$

REMEMBER

What you need to understand about these equations is that e^{ikx} represents plane waves traveling in the $+x$ direction, and e^{-ikx} represents plane waves traveling in the $-x$ direction.

5. **Use the boundary conditions and normalization to find constants.**

   What the Step 4 solutions for $\psi(x)$ mean is that waves can hit the potential step from the left and be either transmitted or reflected. Given that way of looking at the problem, you may note that the wave can be reflected only going to the right, not to the left, so D must equal zero. That makes the wave equation become the following.

   - *Where $x < 0$: $\psi_1(x) = Ae^{ik_1x} + Be^{-ik_1x}$*

   - *Where $x > 0$: $\psi_2(x) = Ce^{ik_2x}$*

REMEMBER

It's useful here to think about what these different terms mean. Remember that a quantum wave function is also a superposition of individual wave functions, and that's what the structure of $\psi_1(x)$ demonstrates. Because it represents $x < 0$, it is the superposition of both the incident wave and the reflected wave. The term Ae^{ik_1x} represents the incident wave, while Be^{-ik_1x} is the reflected wave. For $x > 0$, the wave function $\psi_2(x) = Ce^{ik_2x}$ is the transmitted wave, because that represents the portion that makes it to the other side of the potential step.

To find the values of these constants, I introduce you to some new mathematical tools, which you use in the calculations shown in the following two sections.

Calculating the probability of reflection or transmission

You can calculate the probability that the particle will be reflected or transmitted through the potential step by calculating the reflection and transmission coefficients. The *reflection coefficient* represents how much of the wave is reflected, while the *transmission coefficient* represents how much of the wave is transmitted. As with many things in quantum physics, these ideas from classical physics take on a strange perspective in the quantum realm.

TIP

The entire incident wave is either reflected or transmitted. This means that the total magnitude of the incident wave equals the sum of the reflected and transmitted waves. And so, the reflection coefficient plus the transmission coefficient must equal 1.

You can derive these coefficients from the *current density* J(x), which, in terms of the wave function, looks like the following equation.

$$J(x) = \frac{i\hbar}{2m}\left[\psi(x)\frac{d\psi^*(x)}{dx} - \psi^*(x)\frac{d\psi(x)}{dx}\right]$$

If J_r is the reflected current density, and J_i is the incident current density, then R, the reflection coefficient, and T, the transmission coefficient, are

$$R = \frac{J_r}{J_i}$$

$$T = \frac{J_t}{J_i}$$

You now have to calculate J_r, J_i, and J_t. Actually, that's not so hard. Start with J_i. Because the incident part of the wave is $\psi_i(x) = Ae^{ik_1x}$, the incident current density is

$$J_i = \frac{i\hbar}{2m}\left[Ae^{ik_1x}\frac{d}{dx}(Ae^{-ik_1x}) - Ae^{-ik_1x}\frac{d}{dx}(Ae^{ik_1x})\right] = \frac{\hbar k_1}{m}|A|^2$$

J_r and J_t work in the same way, yielding

$$J_r = \frac{\hbar k_1}{m}|B|^2$$

$$J_t = \frac{\hbar k_2}{m}|C|^2$$

So, you have this set of equations for the reflection coefficient, R, and transmission coefficient, T:

$$R = \frac{J_r}{J_i} = \frac{|B|^2}{|A|^2}$$

$$T = \frac{J_t}{J_i} = \frac{k_2|C|^2}{k_1|A|^2}$$

Those pesky constants: Finding A, B, and C

So how do you figure out the constants A, B, and C? You do that as you figure out the coefficients with the infinite square well potential — with boundary conditions (see the earlier section "Trapping Particles in Infinite Square Well Potentials"). You can't necessarily say that $\psi(x)$ goes to zero, because the potential is no longer infinite. Instead, the boundary conditions are that $\psi(x)$ and $d\psi(x)/dx$ are continuous across the potential step's boundary. In equation form, you have

» $\psi_1(0) = \psi_2(0)$

» $\frac{d\psi_1}{dx}(0) = \frac{d\psi_2}{dx}(0)$

And you know the following:

» **Where x < 0:** $\psi_1(x) = Ae^{ik_1x} + Be^{-ik_1x}$

» **Where x > 0:** $\psi_2(x) = Ce^{ik_2x}$

So, when you plug in the terms, and apply some algebra and calculus:

$$\psi_1(0) = \psi_2(0) \Rightarrow Ae^{ik_1 0} + Be^{ik_1 0} = Ce^{ik_2 0} \Rightarrow A + B = C$$

$$\frac{d\psi_1}{dx}(0) = \frac{d\psi_2}{dx}(0) \Rightarrow \frac{d}{dx}\left(Ae^{ik_1x} + Be^{-ik_1x}\right) = \frac{d}{dx}Ce^{ik_2x} \Rightarrow k_1A - k_1B = k_2C$$

With these two equations, you can solve for both B and C in terms of A.

$$B = \frac{k_1 - k_2}{k_1 + k_2}A$$

$$C = \frac{2k_1}{k_1 + k_2}A$$

You could normalize the function by integrating the probability across all values of x and setting that integral equal to 1. (See the section "Normalizing the wave function," earlier in the chapter, for more on this calculation.) You could, but you don't need to, so I give you a pass on it this time.

You don't need to normalize the function here because the two coefficients are ratios, and the A value will drop out of them. In particular, you have these equations for reflection and transmission coefficients:

$$R = \frac{|B|^2}{|A|^2} = \frac{(k_1 - k_2)^2}{(k_1 + k_2)^2}$$

$$T = \frac{k_2 |C|^2}{k_1 |A|^2} = \frac{4 k_1 k_2}{(k_1 + k_2)^2}$$

REMEMBER

That's an interesting result, and it disagrees with classical physics, which says that no particle reflection should exist. But when you look at the reflection coefficient ratio, if $k_1 \neq k_2$, then the numerator isn't 0, and particle reflection can happen. Already, you have a result that differs from the classical. The particle can be reflected at the potential step. That's the wavelike behavior of the particle coming into play again.

Consider what happens as V_0 approaches zero. That means that k_1 gets closer to k_2, and R goes to 0 while T goes to 1. A smaller potential energy step reduces the probability of a reflection and increases the probability of a transmission. When $V_0 = 0$, then there actually isn't a potential step and $k_1 = k_2$. You can do the calculations to see what ratios get reflected and transmitted in that case.

Assuming the particle doesn't have enough energy

Okay, now try the case where $E < V_0$ when there's a potential step, as shown in Figure 8-6. In this case, the particle doesn't have enough energy to make it into the region $x > 0$, according to classical physics. See what quantum physics has to say about it.

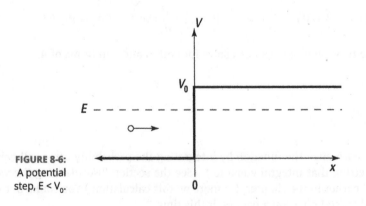

FIGURE 8-6:
A potential step, $E < V_0$.

Again, you start your calculation by setting up your coordinates (see Figure 8-6) and structuring the equations much the same way as shown in the section "Assuming the particle has plenty of energy," earlier in the chapter.

1. **Determine your coordinates to write the Schrödinger equation.**

2. **Apply specific constraints; simplify if possible.**

3. **Restructure the Schrödinger equation into a solvable form.**

   Tackle the region $x < 0$ first. There, the Schrödinger equation would look like this:

   $$\frac{d^2\psi_1}{dx^2}(x) + k_1^2\psi_1(x) = 0$$

   where $k_1^2 = \dfrac{2mE}{\hbar^2}$

   At the risk of jumping ahead a bit to Step 4, you know the solution to this from the previous section's discussion on potential steps, which was mathematically identical:

   $$\psi_1(x) = Ae^{ik_1x} + Be^{-ik_1x} \qquad x < 0$$

   Okay, but what about the region where $x > 0$? That's a different story, because the energy relationships are different than they were in the earlier situation. Here's the Schrödinger equation from that earlier situation:

   $$\frac{d^2\psi_2(x)}{dx^2} + k^2\psi_2(x) = 0 \qquad \text{(where } x > 0\text{)}$$

   where $k^2 = \dfrac{2m(E - V_0)}{\hbar^2}$

   But hang on! This time, $E - V_0$ is less than zero, which would make k imaginary, which is impossible physically. But $V_0 - E$ would be positive, and thus physically possible. In Step 3 from the bullet list in the section "Finding the wave function equation," earlier in the chapter, I defined k^2 for the sake of convenience — largely to tidy up a lot of constants from the equation. The term has served well with that plus sign in front of it. But no rule says that k^2 has to be a plus, so change the sign in the Schrödinger equation from plus to minus:

   $$\frac{d^2\psi_2}{dx^2}(x) - k^2\psi_2(x) = 0 \qquad x > 0$$

   where $k_2^2 = \dfrac{2m(V_0 - E)}{\hbar^2}$

   REMEMBER

   For this changing-sign maneuver to work, consider the special circumstances here — a situation where $E < V_0$. Also keep in mind throughout that these equations only apply for $x > 0$. But that's okay because you already know the solution for $x < 0$.

4. Solve the differential equation for ψ.

Now that the Schrödinger equation is in a solvable form, you solve it. Here are the two linearly independent solutions:

- $\psi(x) = Ce^{-k_2 x}$

- $\psi(x) = De^{k_2 x}$

You add these two solutions together to find the general solution for the wave function in this region:

$$\psi_2(x) = Ce^{-k_2 x} + De^{k_2 x} \qquad x > 0$$

5. Use the boundary condition and normalization to find constants.

Wave functions must be finite everywhere (see the earlier section "Normalizing the wave function"). The second term is clearly not finite as x goes to infinity, so D must equal zero. (The first term would also diverge if x goes to negative infinity, but because the potential step is limited to $x > 0$, that isn't a problem.) Therefore, here's the solution for $x > 0$:

$$\psi_2(x) = Ce^{-k_2 x} \qquad x > 0$$

So, your wave functions for the two regions are

$$\psi_1(x) = Ae^{ik_1 x} + Be^{-ik_1 x} \qquad x < 0$$
$$\psi_2(x) = Ce^{-k_2 x} \qquad x > 0$$

In this case, you can also break these two wave functions into three wave functions that represent the different possible parts of the wave function.

» Incident wave function: $\psi_i(x) = Ae^{ik_1 x}$

» Reflected wave function: $\psi_r(x) = Be^{-ik_1 x}$

» Transmitted wave function: $\psi_t(x) = Ce^{-k_2 x}$

Finding transmission and reflection coefficients

Now you can figure out the reflection and transmission coefficients, R and T. The case $E > V_0$ is calculated by taking ratios in the section "Assuming the particle has plenty of energy," earlier in the chapter. You could do all that work again, but in this situation, you can take a shortcut.

The transmitted wave function, $\psi_t(x) = Ce^{-k_2 x}$ (from the preceding section), doesn't contain i, so $\psi_t(x)$ is completely real. Here's a quick reminder of the transmitted current density function:

$$J_t = \frac{i\hbar}{2m}\left[\psi_t(x)\frac{d\psi_t^*(x)}{dx} - \psi_t^*(x)\frac{d\psi_t(x)}{dx}\right]$$

Since $\psi_t(x)$ is real, that means the complex conjugate will just be the same as the function itself, resulting in

$$J_t = \frac{i\hbar}{2m}\left[\psi_t(x)\frac{d\psi_t(x)}{dx} - \psi_t(x)\frac{d\psi_t(x)}{dx}\right]$$

Conveniently, the term inside the brackets is equal to zero, so $J_t = 0$. The transmission coefficient, T, is based on J_t (as outlined in the section "Assuming the particle has plenty of energy," earlier in the chapter), and so you know that T = 0. Because the sum of the translation coefficients R + T = 1, you also know that R = 1. None of the wave is being transmitted. The wave is entirely reflected, just as in the classical solution.

The nonzero solution: Finding a particle in x > 0

Despite the complete reflection of the wave, a significant difference does exist between the quantum and classical solutions. In the quantum solution, you actually have a nonzero chance of finding the particle in the region x > 0. The probability density for x > 0 (representing the probability in this region) is

$$P(x) = |\psi_t(x)|^2$$

Plugging in for the transmitted wave function $\psi_t(x)$ gives you

$$P(x) = |\psi_t(x)|^2 = |C|^2 e^{-2k_2 x}$$

You can use the following boundary conditions at x = 0 (where the two wave functions coincide with each other) to solve for C in terms of A:

» $\psi_1(0) = \psi_2(0)$

» $\dfrac{d\psi_1}{dx}(0) = \dfrac{d\psi_2}{dx}(0)$

Using the boundary conditions gives you the following equation:

$$P(x) = |C|^2 e^{-2k_2 x} = \frac{4k_1^2 |A| e^{-2k_2 x}}{k_1^2 + k_2^2}$$

The probability density does fall quickly to zero as x gets large, but near x = 0, it has a nonzero value. Figure 8-7 shows what the probability density looks like for the $E < V_0$ case of a potential step.

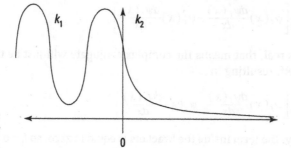

FIGURE 8-7:
The value of k by
region, $E < V_0$.

Tunneling through Forbidden Regions

In the previous section, I describe a difference between classical and quantum mechanical solutions that relates to where you might find transmitted particles. For example, you have the potential for finding particles in classically forbidden regions. Quantum mechanically, the phenomenon where particles can get through regions that they're classically forbidden to enter is called *quantum tunneling*. Tunneling is possible because, in quantum mechanics, particles show wave properties.

Tunneling is one of the most exciting results of quantum physics, and it happens because of the spread in the particles' wave functions. This is, of course, a microscopic effect — so don't try to walk through any closed doors — but it's a significant one. Among other effects, tunneling makes it possible to build working transistors and integrated circuits.

Hitting the Wall: Particles and Potential Barriers

What if the particle could work its way through a potential step — that is, the step didn't extend out to infinity but had a limited extent? Then you'd have a *potential barrier*, which is set up something like this:

» $V(x) = 0$, where $x < 0$ or $x > a$

» $V(x) = V_0$, where $0 \leq x \leq a$

Figure 8-8 illustrates how this potential barrier looks.

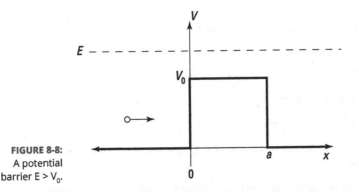

FIGURE 8-8:
A potential
barrier $E > V_0$.

When solving the Schrödinger equation for a potential barrier, you have to consider two cases that correspond to the situations in which the particle has more energy or less energy than the potential barrier. In other words, if E is the energy of the incident particle, the two cases to consider are $E > V_0$ and $E < V_0$.

Getting through potential barriers when $E > V_0$

You can apply the same steps that you find in the section "Finding the wave function equation," earlier in this chapter, to construct and solve the Schrödinger equation for its two cases. In this section, the equations are for the case in which $E > V_0$.

1. **Determine your coordinates to write the Schrödinger equation.**

2. **Apply specific constraints; simplify if possible.**

3. **Restructure the Schrödinger equation into a solvable form.**

   In this case (where $E > V_0$), the particle has enough energy to pass through the potential barrier and end up in the $x > a$ region. This is what the Schrödinger equation looks like:

   - For the region $x < 0$: $\dfrac{d^2\psi_1}{dx^2}(x) + k_1^2\psi_1(x) = 0$ where $k_1^2 = \dfrac{2mE}{\hbar^2}$

   - For the region $0 \le x \le a$: $\dfrac{d^2\psi_2}{dx^2}(x) + k_2^2\psi_2(x) = 0$ where $k_2^2 = \dfrac{2m(E - V_0)}{\hbar^2}$

   - For the region $x > a$: $\dfrac{d^2\psi_3}{dx^2}(x) + k_1^2\psi_3(x) = 0$ where $k_1^2 = \dfrac{2mE}{\hbar^2}$

4. Solve the differential equations for ψ.

The solutions for $\psi_1(x)$, $\psi_2(x)$, and $\psi_3(x)$ are the following:

- *where x < 0:* $\psi_1(x) = Ae^{ik_1x} + Be^{-ik_1x}$
- *where 0 ≤ x ≤ a:* $\psi_2(x) = Ce^{ik_2x} + De^{-ik_2x}$
- *where x > a:* $\psi_3(x) = Ee^{ik_1x} + Fe^{-ik_1x}$

In fact, because there's no leftward-traveling wave in the x > a region, F = 0, so $\psi_3(x) = Ee^{ik_1x}$.

5. Use the boundary conditions and normalization to find constants.

To determine A, B, C, D, and E, you use the boundary conditions, at x = 0 and x = a, which work out here to be the following:

$$\psi_1(0) = \psi_2(0)$$

$$\frac{d\psi_1}{dx}(0) = \frac{d\psi_2}{dx}(0)$$

$$\psi_2(a) = \psi_3(a)$$

$$\frac{d\psi_2}{dx}(a) = \frac{d\psi_3}{dx}(a)$$

From these equations, you get the following:

- $A + B = C + D$
- $ik_1(A - B) = ik_2(C - D)$
- $Ce^{ik_2a} + De^{-ik_2a} = Ee^{ik_1a}$
- $ik_2Ce^{ik_2a} - ik_2De^{-ik_2a} = ik_1Ee^{ik_1a}$

By putting all of these equations together, you get this equation for the coefficient E (this does *not* represent energy!) in terms of A:

$$E = 4k_1k_2Ae^{-ik_1a}\left[4k_1k_2\cos(k_2a) - 2i\left(k_1^2 + k_2^2\right)\sin(k_2a)\right]^{-1}$$

Now determine the transmission coefficient, T, which works out to be

$$T = \frac{|E|^2}{|A|^2} = \left[1 + \frac{1}{4}\left(\frac{k_1^2 - k_2^2}{k_1k_2}\right)^2 \sin^2(k_2a)\right]^{-1}$$

REMEMBER

Note that as k_1 goes to k_2, T goes to 1, which is what you'd expect. (See the end of the earlier section "Those pesky constants: Finding A, B, and C" for a discussion of these expectations.)

So, how about R, the reflection coefficient? I'll spare you the algebra, but here's what R equals:

$$R = \frac{\left(k_2^2 - k_1^2\right)\sin^2\left(k_2 a\right)}{4k_1^2 k_2^2 + \left(k_1^2 - k_2^2\right)^2 \sin^2\left(k_2 a\right)}$$

And Figure 8-9 illustrates the probability density, $|\psi(x)|^2$, for the potential barrier when $E > V_0$.

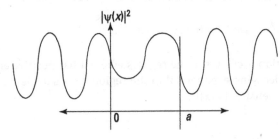

FIGURE 8-9:
$|\psi(x)|^2$ for a
potential barrier
$E > V_0$.

Tunneling through: Potential barriers when $E < V_0$

What happens if the particle doesn't have as much energy as the potential of the barrier? In other words, you're now facing the situation shown in Figure 8-10.

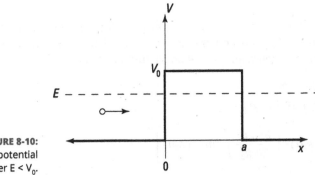

FIGURE 8-10:
A potential
barrier $E < V_0$.

Most of the work for this problem is identical to the work in the preceding section, except in the region of the barrier itself, $0 \leq x \leq a$. The Schrödinger equation in Step 3 (from the previous section) for that region looks like this:

$$\frac{d^2 \psi_2(x)}{dx^2} - k_2^2 \psi_2(x) = 0$$

where $k_2^2 = \dfrac{2m(V_0 - E)}{\hbar^2}$

Since $E - V_0$ is less than 0, which would make k imaginary, you have to perform the trick of flipping them and making the second term negative to get a non-imaginary equation (see the section "Assuming the particle doesn't have enough energy").

The solutions for $\psi_1(x)$ and $\psi_2(x)$ will be identical to the preceding section, but now the solution to this differential equation is real, not imaginary:

$$\psi_2(x) = Ce^{k_2 x} + De^{-k_2 x} \text{ where } 0 \leq x \leq a$$

The wave function oscillates in the regions where it has positive energy, $x < 0$ and $x > a$, but is a decaying exponential in the region $0 \leq x \leq a$. Figure 8-11 illustrates the probability density, $|\psi(x)|^2$.

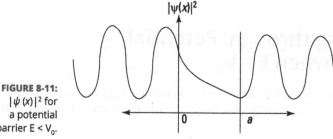

FIGURE 8-11:
$|\psi(x)|^2$ for
a potential
barrier $E < V_0$.

As you may expect, you use the same continuity conditions to determine A, B, and E. How about the reflection and transmission coefficients, R and T? Skipping through a fair bit of algebra and trig, R and T turn out to be:

$$R = \frac{4k_1^2 - k_2^2}{\left(k_1^2 + k_2^2\right)^2 \sinh(k_2 a)}$$

$$T = \left[\cosh(k_2 a) + \left(\frac{k_1^2 - k_2^2}{2k_1^2 k_2^2}\right)^2 \sinh^2(k_2 a)\right]^{-1}$$

Despite the equation's complexity, it's amazing that the expression for T can be nonzero. Classically, particles without enough energy can't enter the forbidden zone $0 \leq x \leq a$ because their energy is less than the potential. In quantum physics, they can! I explain this weird quantum quirk in the section "Tunneling through Forbidden Regions," earlier in the chapter.

Particles Unbound: Solving the Schrödinger Equation for Free Particles

In the previous examples in this chapter, I focus on particles that are constrained or bound in some way, engaging with a potential energy that is influencing their behavior. It's time to let those particles slip off the leash! What about particles outside any square well — that is, free particles? You can find plenty of particles that act freely in the universe, and quantum physics has something to say about them.

1. **Determine your coordinates to write the Schrödinger equation.**

   Since this free particle can be anywhere in the universe, the coordinates don't have to be defined in any specific orientation. Here's the general form of the Schrödinger equation in one dimension:

   $$\frac{-\hbar^2}{2m}\frac{d^2}{dx^2}\psi(x)+V(x)\psi(x)=E\psi(x)$$

2. **Apply specific constraints; simplify if possible.**

3. **Restructure the Schrödinger equation into a solvable form.**

   A free particle doesn't have any potential energy acting on it, so it's in a region of space where V(x) = 0. Applying that constraint to the equation in Step 1, and moving the energy term over to the lefthand side, gives

   $$\frac{-\hbar^2}{2m}\frac{d^2}{dx^2}\psi(x)-E\psi(x)=0$$

   And you can rewrite this as

   $$\frac{d^2}{dx^2}\psi(x)+k^2\psi(x)=0$$

   where the wave number, k, is $k^2=\frac{2mE}{\hbar^2}$.

4. **Solve the differential equation for ψ.**

   You can write the general solution to this Schrödinger equation as

   $\psi(x) = Ae^{ikx} + Be^{-ikx}$

   And you can add the time-dependent elements to the equation, to get this time-dependent wave function:

   $$\psi(x,\ t)=A\exp\left(ikx-\frac{iEt}{\hbar}\right)+B\exp\left(-ikx-\frac{iEt}{\hbar}\right)$$

5. **Use the boundary conditions and normalization to find constants.**

Although setting up the equation and solving it can be technically complex, the real challenge always seems to come at the final stage that involves defining constants in your specific situation. That's no different here!

TIP

Because you have a free particle in space, no boundary conditions exist for you to apply. This situation is distinctly different from the chapter's earlier examples. In this case, you have to apply normalization. The particle must have a probability of 1 of existing somewhere in space, and you'd find that by integrating the probability density of the wave function, across all of space.

Unfortunately, integrating across all of space is going to result in terms diverging (and some other problems, as described in the nearby sidebar, "Heisenberg Uncertainty Problem"). Normalizing this particular function, without any boundary conditions to limit it, is going to again require some new tools, which I introduce in the remainder of this chapter.

Getting a physical particle with a wave packet

TIP

If you have a number of solutions to the Schrödinger equation, any linear combination of those solutions is also a solution. So that's the key to getting a physical particle: You add various wave functions together so that you get a wave packet, which is a collection of wave functions of the form $e^{i(kx - Et/\hbar)}$. These wave functions interfere constructively at one location and interfere destructively (go to zero) at all other locations:

$$\psi(x, t) = \sum_{n-1}^{\infty} \phi_n e^{i\left(\frac{kx-Et}{\hbar}\right)}$$

This is usually written as a continuous integral:

$$\psi(x, t) = \frac{1}{(2\pi)^{\frac{1}{2}}} \int_{-\infty}^{+\infty} \phi(k, t) e^{i\left(\frac{kx-Et}{\hbar}\right)} dk$$

The term $\phi(k, t)$ is the amplitude of each component wave function, and you can find $\phi(k, t)$ from the Fourier transform of the equation:

$$\phi(k, t) = \frac{1}{(2\pi\hbar)^{\frac{1}{2}}} \int_{-\infty}^{+\infty} \psi(x, t) e^{-i\left(\frac{kx-Et}{\hbar}\right)} dx$$

Because $k = p/\hbar$, you can also write the wave packet equations like this, in terms of p, not k:

$$\psi(x, t) = \frac{1}{(2\pi\hbar)^{\frac{1}{2}}} \int_{-\infty}^{+\infty} \phi(p, t) e^{i\left(\frac{px-Et}{\hbar}\right)} dp$$

$$\phi(p, t) = \frac{1}{(2\pi\hbar)^{\frac{1}{2}}} \int_{-\infty}^{+\infty} \psi(x, t) e^{-i\left(\frac{px-Et}{\hbar}\right)} dx$$

Well, you may be asking yourself just what's going on here. It looks like $\psi(x, t)$ is defined in terms of $\phi(p, t)$, but $\phi(p, t)$ is defined in terms of $\psi(x, t)$. That looks pretty circular.

The answer is that the two previous equations aren't definitions of $\psi(x, t)$ or $\phi(p, t)$; they're just equations relating the two. You're free to choose your own wave packet shape — for example, you may specify the shape of $\phi(p, t)$, and $\psi(x, t) = \frac{1}{(2\pi)^{1/2}} \int_{-\infty}^{+\infty} \phi(k, t) e^{i(kx-Et/\hbar)} dk$ would let you find $\psi(x, t)$.

Going through a Gaussian example

Here's an example in which you get concrete, selecting an actual wave packet shape. Choose a so-called Gaussian wave packet — illustrated in Figure 8-12 — which is localized in one place, and zero in the others.

The amplitude $\phi(k)$ you may choose for this wave packet is

$$\phi(k) = A \exp\left[\frac{-a^2(k-k_0)^2}{4}\right]$$

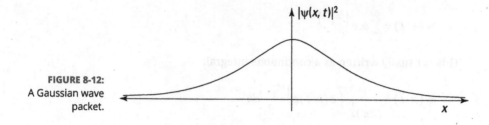

FIGURE 8-12:
A Gaussian wave
packet.

You start by normalizing $\phi(k)$ to determine what A is. Here's how that works:

$$1 = \int_{-\infty}^{+\infty} |\phi(k)|^2 \, dk$$

Substituting in $\phi(k)$ gives you this equation:

$$1 = |A|^2 \left| \int_{-\infty}^{+\infty} \exp\left[\frac{-a^2}{2}\right](k-k_0)^2 \right| dk$$

Doing the integral (that means looking it up in math tables) gives you the following:

$$1 = |A|^2 \left| \left[\frac{2\pi}{a^2}\right]^{\frac{1}{2}} \right|$$

Therefore, $A = \left| \left[\frac{a^2}{2\pi}\right]^{\frac{1}{4}} \right|$.

So here's your wave function:

$$\psi(x) = \frac{1}{(2\pi)^{\frac{1}{2}}} \left[\frac{a^2}{2\pi}\right]^{\frac{1}{4}} \left| \int_{-\infty}^{+\infty} \exp\left[\frac{-a^2(k-k_0)^2}{4}\right] \right| e^{ikx} dk$$

You can evaluate this little gem of an integral to find the following:

$$\psi(x) = \left[\frac{2}{\pi a^2}\right]^{\frac{1}{4}} \exp\left[\frac{-x^2}{a^2}\right] e^{ik_0 x}$$

So, that's the wave function for this Gaussian wave packet (*Note:* The $\exp[-x^2/a^2]$ is the Gaussian part that gives the wave packet the distinctive shape that you see in Figure 8-12) — and it's already normalized.

Want to take the wave function out for a test? Remember the main function of the wave function is that it relates to the probability density of finding a particle in a given location. Now you can use this wave packet function to determine the probability that the particle will be in, say, the region $0 \le x \le a/2$. The probability is

$$\int_0^{a/2} |\psi(x)|^2 \, dx = \sqrt{\frac{2}{\pi a^2}} \int_0^{a/2} \exp\left(\frac{-2x^2}{a^2}\right) dx$$

$$= \frac{1}{3}$$

So, the probability that the particle will be in the region $0 \le x \le a/2$ is $1/3$, which is roughly the same probability as being one standard deviation from the mean in a Gaussian distribution. So, the distance of $a/2$ is roughly a standard deviation in this case. Cool! (It's almost like it wasn't chosen at random.)

Chapter **9**

Back and Forth with Harmonic Oscillators

H armonic oscillators are physics setups with periodic motion, such as things bouncing on springs or tick-tocking on pendulums. You're probably already familiar with harmonic oscillator problems in the macroscopic arena, but now you're going microscopic. You can approximate many physical cases — such as atoms in a crystal structure — by using harmonic oscillators.

In this chapter, you see exact solutions to harmonic oscillator problems as well as computational methods for solving them. I show you how harmonic oscillators can be interpreted in different energy states and solve them using the Dirac notation with bras and kets.

Note: Before diving into this chapter, you may want to look over Chapter 2, which covers classical periodic motion and also some key features of trigonometry. This chapter assumes familiarity with those concepts, as well as calculus, since I use calculus to transform some periodic motion equations. Working with the state equations uses knowledge of linear algebra. In addition, to solve the Schrödinger equation (discussed in Chapter 7), you need a familiarity with differential equations.

Grappling with the Harmonic Oscillator Hamiltonians

Okay, time to start talking Hamiltonians (and I'm not referring to fans of the U.S. Founding Father, Alexander Hamilton). The *Hamiltonian* is an operator that helps you find the energy levels of a system.

Going classical with harmonic oscillation

You may remember two primary harmonic motion cases from classical physics (and Chapter 2): the simple pendulum and a mass suspended from a spring. Analogous harmonic relationships also apply to interpreting relationships between charge, current, voltage, and other electrical properties. And of course, the periodic motion of waves shows up in fluid dynamics and acoustics.

Because of the recurrence of this mathematical behavior, harmonic oscillations are incredibly well-studied in classical physics, and I can use that understanding as the springboard for presenting the quantum harmonic oscillators.

So, what do we know from classical physics? The force on an object in harmonic oscillation is defined by *Hooke's Law*:

$$F = -kx$$

In this equation, k is the spring constant, measured in Newtons/meter, and x is displacement, in meters. The spring constant is a feature of the physical structure that is oscillating. In the case of a spring, it represents how resistant that spring is to being stretched. A loose spring will have a low value of k, while a very tight spring will have a larger value of k, because the force needed to stretch a loose spring is less than the force needed to stretch a tight spring.

REMEMBER

The key point here is that the restoring force on whatever is in harmonic motion is proportional to its displacement. In other words, the farther you stretch a spring, the harder it pulls back when released.

Because $F = ma$ (per Newton's second law of motion, as discussed in Chapter 2), where m is the mass of the particle in harmonic motion and a is its instantaneous acceleration, you can substitute for F and write this equation as

$$ma + kx = 0$$

Time now to apply some calculus, because instantaneous acceleration (a) is the second derivative of displacement (x) with respect to time (t):

$$a = \frac{d^2x}{dt^2}$$

So, substituting for a, you can rewrite the force equation from Newton's second law of motion as

$$ma + kx = m\frac{d^2x}{dt^2} + kx = 0$$

Dividing by the mass of the particle gives you the following:

$$\frac{d^2x}{dt^2} + \frac{kx}{m} = \frac{0}{m}$$

Now you can do a little substitution to reduce the number of variables you need to track in the resulting differential equation. If you take $\frac{k}{m} = \omega^2$ (where ω is the angular frequency), this becomes

$$\frac{d^2x}{dt^2} + \omega^2 x = 0$$

Through the wonders of calculus, you can solve this second-order differential equation for x, where A and B are constants:

$$x = A\sin\omega t + B\cos\omega t$$

The solution is an oscillating one because it involves sines and cosines, which represent periodic waveforms. You get this far without even stepping into the microscopic realm where you must begin applying the quantum rules.

Understanding total energy in quantum oscillation

Now look at harmonic oscillators in quantum physics terms. The related concepts don't represent pendulums or masses on springs, but instead, a *quantum oscillator* is any quantum system having movement in which the force is proportional to the amount of displacement from a stable equilibrium point. It turns out that you can model a lot of microscopic systems in this way, so the quantum oscillator system becomes one of the most important models in all of quantum physics!

Putting the Hamiltonian to use

The Hamiltonian (H) operator (see Chapter 7 where I introduce the Hamiltonian operator) yields the *total energy of a system*, which means that it gives the sum of kinetic energy (KE) and potential energy (PE), so you can break the Hamiltonian into those two components: H = KE + PE.

For a harmonic oscillator, here's what these energies are equal to.

>> **The kinetic energy at any one moment,** where p is the particle's momentum and m is its mass:

$$KE = \frac{p^2}{2m}$$

>> **The particle's potential energy,** where k is the spring constant, x is displacement, and the angular momentum is $\omega^2 = \frac{k}{m}$, gives you

$$PE = \frac{1}{2}kx^2 = \frac{1}{2}m\omega^2 x^2$$

In quantum physics terms, you use operators on matrices of values instead of individual variables. This means that you can write the Hamiltonian by adding the kinetic energy and potential energy together, giving you

$$H = \frac{P^2}{2m} + \frac{1}{2}m\omega^2 X^2$$

where P and X are the momentum and position operators.

Applying the Hamiltonian to eigenstates

You can apply the Hamiltonian operator to various *eigenstates* (where one of the system variables has a determinate fixed value — see Chapter 7 for more on eigenstates), $|\psi\rangle$, of the harmonic oscillator to get the total energy, E, of those eigenstates:

$$H|\psi\rangle = \frac{P^2}{2m}|\psi\rangle + \frac{1}{2}m\omega^2 X^2|\psi\rangle = E|\psi\rangle$$

The problem now becomes one of finding the eigenstates and eigenvalues. However, this doesn't turn out to be an easy task. Unlike the potential V(x) covered in Chapter 8, V(x) for a harmonic oscillator is more complex, depending as it does on x^2. The reason to start with potential wells (in Chapter 8) is because they are particularly straightforward!

You have to be clever to make these harmonic oscillator equations solvable. The way you solve harmonic oscillator problems in quantum physics is with operator algebra — that is, you introduce a new set of operators. And they're coming up in the next section.

Creation and Annihilation: Introducing the Harmonic Oscillator Operators

Quantum physics was basically built as a bunch of mathematical tricks for working with matrices representing quantum values. You could transform those sets of values and get numbers out that you could then test against experimental results. For over a century, experiments have repeatedly confirmed the results of these mathematical tricks, which suggests that these tricks work.

One useful mathematical trick that works well in quantum physics enables you to modify the tougher Hamiltonian equation into a more manageable form. This trick comes by introducing two operators with monumental-sounding names:

» **The creation operator ($a^\dagger$)** raises the energy of an eigenstate by one level. So, for example, if the harmonic oscillator is in the fourth energy level, the creation operator raises it to the fifth level.

» **The annihilation operator (a)** lowers the eigenstate's energy by one level (the reverse of the creation operator).

TIP

These creation and annihilation operators enable physicists to solve for the energy spectrum without a lot of work solving for the actual eigenstates. Once you solve for one of the states, you can use these operators to determine the energies in adjacent states.

The easiest state to solve for is usually the ground state, $n = 0$. You might think that the ground state would be a no-energy state of 0, but remember how quantum mechanics works! You don't often deal with completely still objects in the quantum realm. Almost as soon as Werner Heisenberg came up with the matrix mechanics form of quantum mechanics, he realized that uncertainty always exists in the position and momentum of a quantum system. (I cover the Heisenberg uncertainty principle in greater detail in Chapter 3.)

REMEMBER

Nothing ever sits completely still in quantum mechanics, so a quantum harmonic oscillator will always have some level of motion. These quantum gyrations mean that the ground state of energy E_0 at $n = 0$ will never be exactly 0 — although it will be the lowest possible energy state. And that's what makes it the ground state.

Mind Your p's and q's: Getting the Energy State Equations

You can solve for the energy spectrum of a quantum harmonic oscillator by using the following steps.

1. **Define operators _p_ and _q_.**

First, introduce two new operators, _p_ and _q_, which are dimensionless. They relate to the P (momentum) and X (position) operators this way:

- $p = \dfrac{P}{\sqrt{m\hbar\omega}}$

- $q = X\sqrt{\dfrac{m\omega}{\hbar}}$

It is common to use _p_ and _q_ in linear algebra to substitute in transforming equations, and unfortunately, it is confusing when they are used alongside the variables used by physicists for momentum. The _p_ operator _uses_ the momentum operator, P, but the _p_ operator is **not** the momentum operator itself. And it isn't the variable for momentum, either.

2. **Define _a_ and _a_† in terms of _p_ and _q_.**

You use these two new operators, _p_ and _q_, as the basis of the annihilation operator, _a_, and the creation operator, _a_†:

- $a = \dfrac{1}{\sqrt{2}}(q + ip)$

- $a^{\dagger} = \dfrac{1}{\sqrt{2}}(q - ip)$

3. **Rewrite the Hamiltonian in terms of _a_ and _a_†.**

Here is what you get (but I save you all the intervening steps):

$$H = \hbar\omega\left(a^{\dagger}a + \frac{1}{2}\right)$$

4. **Define the number operator, N.**

The quantum physicists went crazy creating new operators here, even giving a name to $a^{\dagger}a$: the N or _number operator_. So here's how you can write the Hamiltonian:

$$H = \hbar\omega\left(N + \frac{1}{2}\right)$$

The N operator returns the number of the energy level of the harmonic oscillator. If you denote the eigenstates of N as $|n\rangle$, you get this, where n is the number of the nth state:

$$N|n\rangle = n|n\rangle$$

5. **Solve for E_n.**

   Because $H = \hbar\omega\left(N + {}^1/_2\right)$, and because $H|n\rangle = E_n|n\rangle$, by comparing the previous two equations, you have

   $$E_n = \left(n + \frac{1}{2}\right)\hbar\omega \quad n = 0,\ 1,\ 2\ldots$$

Amazingly, that gives you the energy eigenvalues of the nth state of a quantum mechanical harmonic oscillator. So here are the energy states:

» **The ground state energy** corresponds to $n = 0$,

$$E_0 = \frac{1}{2}\hbar\omega$$

» **The first excited state** is

$$E_1 = \frac{3}{2}\hbar\omega$$

» **The second excited state** has an energy of

$$E_2 = \frac{5}{2}\hbar\omega$$

And so on. That is, the energy levels are discrete and *nondegenerate* (not shared by any two states). Thus, the energy spectrum is made up of equidistant bands.

Finding the Eigenstates

When you have the *eigenstates* (solutions for the energy; see Chapter 7 to find out all about eigenstates), you can determine the allowable states of a system and the relative probability that the system will be in any of those states.

Applying the commutator of operators (again, see Chapter 7) to a and $a^\dagger$ you get the following:

$$\left[a,\ a^\dagger\right] = \frac{1}{2}[q + ip,\ q - ip]$$

This is equal to the following:

$$[a, a^\dagger] = \frac{1}{2}[q + ip, q - ip] = -i[q, p]$$

This equation breaks down to $[a, a^\dagger] = 1$. And putting together this equation with $H = \hbar\omega\left(N + \frac{1}{2}\right)$, you get $[a, H] = \hbar\omega a$ and $[a^\dagger, H] = -\hbar\omega a^\dagger$.

Finding the energy of $a|n\rangle$

Okay, with the commutator relations, you're ready to go. Follow these steps:

1. **Apply the commutator.**

   Consider a case where the energy of state $|n\rangle$ is E_n. What is the energy of the state $a|n\rangle$? Well, to find this, rearrange the commutator $[a, H] = \hbar\omega a$ to get $Ha = aH - \hbar\omega a$.

   If it is not immediately clear how you go from the commutator to this final form, you can apply the definition of the commutator, $[A, B] = AB - BA$:

   $$[a, H] = \hbar\omega a$$
   $$aH - Ha = \hbar\omega a$$
   $$Ha = aH - \hbar\omega a$$

   Note: This version of the equation is the one that you use in the next step when applying the Hamiltonian to the annihilation operator.

2. **Apply the Hamiltonian.**

   Now use this version to write the Hamiltonian of $a|n\rangle$ and apply the transformation from Step 1, like this:

   $$Ha|n\rangle$$
   $$= (aH - \hbar\omega a)|n\rangle$$
   $$= (E_n - \hbar\omega)a|n\rangle$$

REMEMBER

So $a|n\rangle$ is also an eigenstate of the harmonic oscillator, with energy $E_n - \hbar\omega$, not E_n. That's why a is called the annihilation or lowering operator: It lowers the energy level of a harmonic oscillator eigenstate by one level.

Finding the energy of $a^\dagger|n\rangle$

So, what's the energy level of $a^\dagger|n\rangle$, the creation operator? Follow these steps to find out:

1. **Apply the commutator of the creation operator.**

   For the creation operator, the commutator is $\left[a^\dagger, H\right] = -\hbar\omega a^\dagger$, which is equivalent to $a^\dagger H - Ha^\dagger = -\hbar\omega a^\dagger$ and $Ha^\dagger = a^\dagger H + \hbar\omega a^\dagger$.

2. **Apply the Hamiltonian to the creation operator.**

   Now that you have the equation from Step 1, you apply the Hamiltonian to the creation operator on the state vector:

   $$Ha^\dagger|n\rangle$$
   $$= \left(a^\dagger H + \hbar\omega a^\dagger\right)|n\rangle$$
   $$= (E_n + \hbar\omega)a^\dagger|n\rangle$$

REMEMBER

This means that $a^\dagger|n\rangle$ is an eigenstate of the harmonic oscillator, with energy $E_n + \hbar\omega$, not just E_n — that is, the $a^\dagger$ operator raises the energy level of an eigenstate of the harmonic oscillator by one level.

Deriving a and a† directly

In the preceding section, you find out that $H(a|n\rangle) = (E_n - \hbar\omega)(a|n\rangle)$ and $H\left(a^\dagger|n\rangle\right) = (E_n + \hbar\omega)\left(a^\dagger|n\rangle\right)$. You can derive the annihilation (lowering) and creation (raising) operators directly by using the operators on the current state vector. You get an output that is the new state vector, either lowered (for the annihilation/lowering operator a) or increased (for the creation/raising operator $a^\dagger$).

$$a|n\rangle = C|n-1\rangle$$
$$a^\dagger|n\rangle = D|n+1\rangle$$

C and D are positive constants, but what do they equal? The states $|n-1\rangle$ and $|n+1\rangle$ have to be normalized, which means that $\langle n-1|n-1\rangle = \langle n+1|n+1\rangle = 1$. (See Chapter 8 for information on normalizing a wave function.) So, take a look at the quantity using the C operator:

$$\left(\langle n|a^\dagger\rangle(a|n\rangle\right) = C^2\langle n-1|n-1\rangle$$
$$\left(\langle n|a^\dagger\rangle(a|n\rangle\right) = C^2$$
$$\langle n|a^\dagger a|n\rangle = C^2$$

But you also know that $a^\dagger a = N$, the energy level operator, so you get the following equation:

$$\langle n|N|n\rangle = C^2$$

$N|n\rangle = n|n\rangle$, where n is the energy level, so

$$n\langle n|n\rangle = C^2$$

However, $\langle n|n\rangle = 1$ because it is already normalized, so you can solve for C:

$$n = C^2 \qquad \rightarrow \qquad \sqrt{n} = C$$
$$a|n\rangle = C|n-1\rangle \quad \rightarrow \quad a|n\rangle = \sqrt{n}|n-1\rangle$$

In other words, using the lowering operator, a, on eigenstates of the harmonic oscillator translates into multiplying it by a constant that's defined by energy state.

What about the raising operator, $a^\dagger$? A similar course of reasoning shows that it, too, multiplies by a constant defined by energy state. Here's that equation:

$$a^\dagger|n\rangle = \sqrt{n+1}|n+1\rangle$$

At this point, you know what the energy eigenvalues are and how the raising and lowering operators affect the harmonic oscillator eigenstates. You've made quite a lot of progress, using the a and $a^\dagger$ operators instead of trying to solve the Schrödinger equation.

Finding the harmonic oscillator energy eigenstates

The charm of using the operators a and $a^\dagger$ is that given the ground state, $|0\rangle$, those operators let you find all successive energy states. If you want to find an excited state of a harmonic oscillator, you can start with the ground state, $|0\rangle$, and apply the raising operator, $a^\dagger$. For example, you can do this:

» $|1\rangle = a^\dagger|0\rangle = |1\rangle$

» $|2\rangle = \dfrac{1}{\sqrt{2}} a^\dagger|1\rangle = \dfrac{1}{\sqrt{2!}}\left(a^\dagger\right)^2|0\rangle$

» $|3\rangle = \dfrac{1}{\sqrt{3}} a^\dagger|2\rangle = \dfrac{1}{\sqrt{3!}}\left(a^\dagger\right)^3|0\rangle$

And so on. In general, you have this relation:

$$|n\rangle = \dfrac{1}{\sqrt{n!}}\left(a^\dagger\right)^n|0\rangle$$

Working in position space

This equation is fine as far as it goes, but you want to find $\langle x|0 \rangle = \psi_0(x)$ (the eigenstate of the ground state eigenvector). To do this, refer to Chapter 8 for the wave function equation and follow these steps:

1. **Determine your coordinates to write the Schrödinger equation.**

You already have some equations to work with. These equations use the linear momentum operator, P, and the position operator, X. Since you want to work with a second-order differential equation in terms of the position, x, you want to work entirely in the position space as your coordinate basis.

1a. Write operator q in the position basis.

The q operator is already in the position space (see the section, "Mind Your p's and q's: Getting the Energy State Equations," earlier in the chapter), so no more work is needed there.

1b. Write operator q in the position basis.

The p operator is defined as

$$p = \frac{P}{\sqrt{m\hbar\omega}}$$

Because $P = -i\hbar\frac{d}{dx}$ (as introduced in Chapter 7), you can write

$$p = \frac{-i\hbar}{\sqrt{m\hbar\omega}}\frac{d}{dx} = -i\sqrt{\frac{\hbar}{m\omega}}\frac{d}{dx}$$

And you now have an equation with the p operator entirely in terms of x.

And writing $x_0 = \sqrt{\frac{\hbar}{m\omega}}$, this becomes

$$p = \frac{-i\hbar}{\sqrt{m\hbar\omega}}\frac{d}{dx} = -ix_0\frac{d}{dx}$$

1c. Write operator a in the position basis.

Okay, what about the a operator? You can use the definitions of p and q to get operator a:

$$a = \frac{1}{\sqrt{2}}(q + ip)$$

$$q = X\sqrt{\frac{m\omega}{\hbar}} = X/x_0$$

$$a = \frac{1}{\sqrt{2}}\left(\frac{X}{x_0} + x_0\frac{d}{dx}\right) = \frac{1}{x_0\sqrt{2}}\left(X + x_0^2\frac{d}{dx}\right)$$

1d. Repeat Step 1c for the lowering operator a†.

That turns out to be this:

$$a^{\dagger} = \frac{1}{x_0\sqrt{2}}\left(X - x_0^2\frac{d}{dx}\right)$$

Having both the raising and lowering operators in the position space means you can use them to try to find specific state vectors, starting with the ground state. Using the ground state allows you to write out the Schrödinger equation for the wave function.

2. Apply specific constraints. Simplify if possible.

You want to solve for the ground state in the position space, so you need another one of those clever math tricks that quantum physics is known for.

2a. Use the lowering operator, a, on the ground state.

There is no energy state lower than the ground state, so lowering the ground state has to give you 0. This is written as:

$$a|0\rangle = 0$$

And applying the $\langle x|$ bra gives you $\langle x|a|0\rangle = 0$.

The clever part is that using the lowering operator on the ground state gives you a differential equation that equals zero . . . also known as a *homogeneous* differential equation.

2b. Write the homogeneous differential equation.

Substitute for *a:*

$$\frac{1}{x_0\sqrt{2}}\left(X + x_0^2\frac{d}{dx}\right)\psi_0(x) = 0$$

$$\frac{1}{x_0\sqrt{2}}\left(x\psi_0(x) + x_0^2\frac{d\psi_0(x)}{dx}\right) = 0$$

And you now have an initial form of the Schrödinger equation to work from in Steps 3 and 4.

3. Restructure the Schrödinger equation into a solvable form.

4. Solve the differential equation for ψ.

Multiplying both sides by $x_0\sqrt{2}$, and solving the resulting differential equation results in

$$x\psi_0(x) + x_0^2\frac{d\psi_0(x)}{dx} = 0$$

$$\frac{d\psi_0(x)}{dx} = \frac{-x\psi_0(x)}{x_0^2}$$

$$\psi_0(x) = A\exp\left(-x^2\big/2x_0^2\right)$$

The resulting equation represents a Gaussian function, so the ground state of a quantum mechanical harmonic oscillator is a Gaussian curve, as shown in Figure 9-1. This makes sense, because it is the simplest oscillation state of the system, with only a single oscillation. Higher energy states will have more complex oscillations. (You can find out more about Gaussian functions in the last section of Chapter 8.)

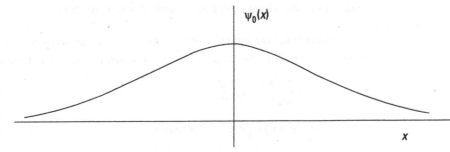

$\psi_0(x)$

x

FIGURE 9-1:
The ground state
of a quantum
mechanical
harmonic
oscillator.

5. Use the boundary condition and normalization to find constants.

One quantity of interest in the ground state equation from Step 4 would be the amplitude, A. Because wave functions must be normalized, set the integral that represents probability equal to 1 and then go through the steps to solve it for A.

$$1 = \int_{-\infty}^{\infty} |\psi_0(x)|^2 dx$$

$$1 = \int_{-\infty}^{\infty} \left| A \exp\left(-x^2 \big/ 2x_0{}^2 \right) \right|^2 dx$$

$$1 = A^2 \int_{-\infty}^{\infty} \left(\exp\left(-x^2 \big/ 2x_0{}^2 \right) \right)^2 dx$$

$$1 = A^2 \int_{-\infty}^{\infty} \exp\left(-x^2 \big/ x_0{}^2 \right) dx$$

$$1 = A^2 \sqrt{\pi} x_0$$

$$A = \frac{1}{\sqrt[4]{\pi} \sqrt{x_0}}$$

When you substitute A in the homogeneous differential equation, the result means that the exact wave function for the ground state of a quantum mechanical harmonic oscillator is

$$\psi_0(x) = \frac{1}{\sqrt[4]{\pi} \sqrt{x_0}} \exp\left(-x^2 \big/ 2x_0{}^2 \right)$$

A little excitement: Finding the first excited state

Now that you have the ground state (from the previous section), you can build from the ground up to figure out the more excited (and exciting!) states. You'll start with the first excited state, $\psi_1(x)$. For this calculation, you don't have to jump all the way back to solving the wave function. You know the ground state, so you can step up to the first excited state simply by using the raising operator (which you discover in the section "Creation and Annihilation: Introducing the Harmonic Oscillator Operators," earlier in the chapter)!

The wave function from the first excited state can be written as $\psi_1(x) = \langle x|1\rangle$ and $|1\rangle = a^\dagger|0\rangle$, so $\psi_1(x) = \langle x|a^\dagger|0\rangle$. And you know that $a^\dagger$ is the following:

$$a^\dagger = \frac{1}{x_0\sqrt{2}}\left(X - x_0^2 \frac{d}{dx}\right)$$

Therefore, $\psi_1(x) = \langle x|a^\dagger|0\rangle$ becomes

$$\langle x|a^\dagger|0\rangle = \frac{1}{x_0\sqrt{2}}\left\langle x\left|\left(X - x_0^2 \frac{d}{dx}\right)\right|0\right\rangle$$

$$= \frac{1}{x_0\sqrt{2}}\left(X - x_0^2 \frac{d}{dx}\right)\langle x|0\rangle$$

And because $\psi_0(x) = \langle x|0\rangle$, you get the following equation:

$$\psi_1(x) = \frac{1}{x_0\sqrt{2}}\left(X - x_0^2 \frac{d}{dx}\right)\psi_0(x)$$

$$= \frac{1}{x_0\sqrt{2}}\left(x - x_0^2 \frac{(-x)}{x_0^2}\right)\psi_0(x)$$

$$= \frac{\sqrt{2}}{x_0}x\psi_0(x)$$

You also know from calculations in the preceding section:

$$\psi_0(x) = \frac{1}{\sqrt[4]{\pi}\sqrt{x_0}}\exp\left(-x^2\Big/2x_0^2\right)$$

Therefore,

$$\psi_1(x) = \frac{\sqrt{2}}{x_0}x\psi_0(x)$$

$$= \frac{\sqrt{2}}{\sqrt[4]{\pi}\sqrt{x_0^3}}x\exp\left(-x^2\Big/2x_0^2\right)$$

You can see a graph of the first excited state in Figure 9-2, where the curve has one *node* (transition through the x axis). Comparing this to the ground state, you may have an idea of how increasing excited states might look, but you want to check it out to know for sure.

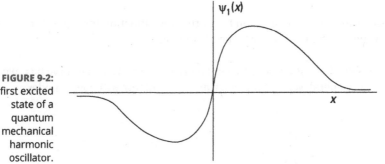

FIGURE 9-2:
The first excited
state of a
quantum
mechanical
harmonic
oscillator.

Finding the second excited state

All right, how about other excited states? Again, you keep repeating the process of applying the raising operator, as in this equation:

$$\psi_2(x) = \frac{1}{\sqrt{2!}} \langle x | (a^\dagger)^2 | 0 \rangle$$

Here I just jump ahead. Substituting for $a^\dagger$, the equation for the wave function of the second excited state becomes

$$\psi_2(x) = \frac{1}{\sqrt{2!}} \frac{1}{2x_0^2} \left(x - x_0^2 \frac{d}{dx} \right) \psi_0(x)$$

Using hermite polynomials to find any excited state

You can generalize the wave function like this:

$$\psi_n(x) = \frac{1}{\sqrt[4]{\pi} \sqrt{2^n n!}} \frac{1}{x_0^{n+1/2}} \left(x - x_0^2 \frac{d}{dx} \right)^n \exp\left(-x^2 \big/ 2x_0^2 \right)$$

To solve this general differential equation, you use hermite polynomials, invented by Simon Laplace in 1810, well before anyone thought of quantum physics. $H_n(x)$ is the nth hermite polynomial. Mathematicians have already done the work of calculating hermite polynomials, so you can look them up in tables of functions. It turns out you can express the wave functions for quantum mechanical harmonic oscillators like this, using the hermite polynomials $H_n(x)$:

$$\psi_n(x) = \frac{1}{\sqrt[4]{\pi} \sqrt{2^n n! x_0}} H_n\left(x \big/ x_0 \right) \exp\left(-x^2 \big/ 2x_0^2 \right)$$

where

$$x_0 = \sqrt{\frac{\hbar}{m\omega}}$$

And you can use this expression to get the wave function for a quantum mechanical harmonic oscillator at any excited state, n.

You can see what $\psi_2(x)$ looks like in Figure 9-3; note that the curve has two nodes here — in general, $\psi_n(x)$ for the harmonic oscillator will have n nodes.

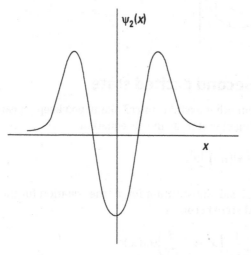

FIGURE 9-3:
The second excited state of a quantum mechanical harmonic oscillator.

Reality check: Putting in some numbers

The preceding section gives you $\psi_n(x)$, and you solve for E_n in the section "Mind Your p's and q's: Getting the Energy State Equations," so you're on top of harmonic oscillators. Now take a look at a real-world example, to see how you can use these equations.

Say that you have a proton undergoing harmonic oscillation with $\omega = 4.58 \times 10^{21} \sec^{-1}$, as shown in Figure 9-4.

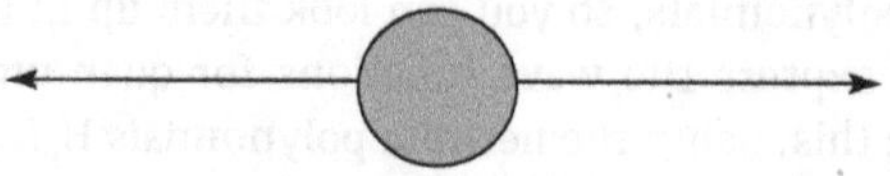

FIGURE 9-4:
A proton undergoing harmonic oscillation.

What are the various energy levels of the proton? You know that in general,

$$E_n = \left(n + \frac{1}{2}\right)\hbar\omega \quad n = 0,\ 1,\ 2\ldots$$

So here are the energies of the proton, in megaelectron-volts (MeV):

» $E_0 = \dfrac{\hbar\omega}{2} = 1.51\,\text{MeV}$

» $E_1 = \dfrac{3\hbar\omega}{2} = 4.52\,\text{MeV}$

» $E_2 = \dfrac{5\hbar\omega}{2} = 7.54\,\text{MeV}$

» $E_3 = \dfrac{7\hbar\omega}{2} = 10.6\,\text{MeV}$

And the pattern continues in this way for subsequent energy levels. Now what about the wave functions? You find the general equation in the last section, look up the values of the hermite polynomials, and then convert all length measurements into femtometers (1 fm $= 1\times10^{-15}$ m). This gives you $x_0 = 3.71$ fm to plug into the equation. Here are the equations for the first few states, where x is measured in femtometers:

» $\psi_0(x) = \dfrac{1}{1.92\sqrt[4]{\pi}}\exp\!\left(-x^2\big/27.5\right)$

» $\psi_1(x) = \dfrac{1}{2.72\sqrt[4]{\pi}}\,2\!\left(x\big/3.71\right)\exp\!\left(-x^2\big/27.5\right)$

» $\psi_2(x) = \dfrac{1}{5.45\sqrt[4]{\pi}}\left[4\!\left(x\big/3.71\right)^2 - 2\right]\exp\!\left(-x^2\big/27.5\right)$

NOTATIONS: WAVE MECHANICS AND MATRIX MECHANICS

The first insights assumed quantum jumps without a clear justification about why they were happening. Niels Bohr tried to build a solid foundation for this new type of physics by gathering some of the world's brightest young physicists in Copenhagen to tackle these problems.

The young Werner Heisenberg, together with his mentor Max Born (1882–1970) and Born's other student Pascual Jordan (1902–1980), realized that the results made sense when viewed as a series of numbers in matrix format.

When the brilliant English physics student Paul Dirac (1902–1984) read Heisenberg's published paper months later, he realized that the matrices presented by Heisenberg were similar to a structure called Poisson brackets. Before the end of 1925, Dirac had independently derived matrix mechanics through another method.

(continued)

(continued)

Physicists then applied these matrix methods to a complicated problem: trying to achieve the same results that Bohr did in his description of electron orbits in an atom. It was a problem too complex for even these brilliant physicists, but fortunately, Heisenberg was a friend of Wolfgang Pauli (1900–1958), who helped them out.

Meanwhile, Bohr's group was still working on a different, calculus-based approach. Erwin Schrödinger (1887–1961) published his wave function, the cornerstone of wave mechanics, in 1926. Many physicists were pleased not to have to learn the new matrix mathematics, but the wave function had problems of its own . . . until Max Born (yes, Heisenberg's mentor) realized that the Schrödinger equation related to quantum probabilities. Dirac would later develop the Dirac notation approach, published in 1939, which is the more streamlined mathematics that we now use for quantum mechanics.

What is the point of this digression into the history of quantum mechanics? The matrix and wave approaches are both equally valid, and some brilliant physicists did a lot of work (and won a whole slew of Nobel Prizes in Physics) to prove that you could jump between them.

Chapter **10**

Working with Angular Momentum on the Quantum Level

I n classical mechanics, you may measure angular momentum by attaching a golf ball to a string and whirling it over your head, which creates an object moving in a circular path. In quantum mechanics, you think in terms of a single molecule made up of two bound atoms rotating around each other. Molecular is the level at which quantum mechanical effects become noticeable. And at that level, it turns out that angular momentum is quantized. And because the quantization has tangible results in many cases — such as the spectrum of excited atoms — it's an important topic. (See Chapter 4 for an introduction to quantum mechanics spectra.)

The equations in this chapter provide the basis for understanding angular momentum in a quantum mechanical system, which has more significance than you might expect. As it turns out, research that went deeper into the behavior of particle physics made amazing discoveries about how angular momentum manifests in these systems. In quantum mechanics, angular momentum reveals a fundamental new property of matter, called *spin* (the subject of Chapter 11).

Note: In this chapter, I'm assuming you have a basic knowledge of classical angular momentum. The letter L represents the angular momentum in the equations used in this chapter, and it relates to the total energy at the quantum level, just like the linear momentum, P, related to it in previous chapters. Also in this chapter, you use the math principles of calculus, linear algebra, and differential equations.

Setting Up the Hamiltonian

Besides having kinetic and potential energy, particles can also have rotational energy. Here's what the Hamiltonian equation (for total energy, see Chapter 9) looks like:

$$H = \frac{L^2}{2I}$$

Here, L is the angular momentum operator and I is the rotation moment of inertia. What are the eigenstates of angular momentum? You can write the following, where ℓ is an eigenvalue of the angular momentum operator L.

$$H|\ell\rangle = \frac{L^2}{2I}|\ell\rangle$$

But that turns out to be incomplete because angular momentum is a vector in three-dimensional space — and it can be pointing in any direction. Angular momentum is typically given by a magnitude and a component in one direction, which is usually the Z direction. So, in addition to the magnitude ℓ, you also specify the component of L in the Z direction, L_z (the choice of Z is arbitrary — you can just as easily use the X or Y direction).

If the quantum number of the Z component of the angular momentum is designated by m (note that this m does *not* represent mass), then the complete eigenstate is given by $|\ell, m\rangle$, so the equation becomes the following:

$$H|\ell, m\rangle = \frac{L^2}{2I}|\ell, m\rangle$$

That's the kind of discussion about eigenstates that I cover in this chapter, and I begin with a discussion of angular momentum.

Ringing the Operators: Round and Round with Angular Momentum

Take a look at Figure 10-1, which depicts a disk rotating in three-dimensional space. Because you're working in three dimensions, you can't simplify the process by ignoring the direction of the vector. Vectors in these cases need to represent both magnitude and direction, a condition which I've largely avoided in previous cases.

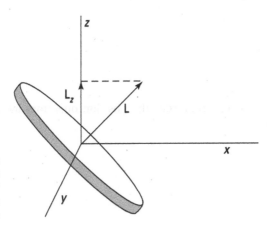

FIGURE 10-1:
A rotating disk with angular momentum vector L.

TIP

The disk's angular momentum vector, **L**, points perpendicular to the plane of rotation. Here, you can apply the right-hand rule: If you hold your right hand so that you can curve your four fingers in the direction something is rotating, your extended thumb points in the direction of the **L** vector.

Having the **L** vector point out of the plane of rotation has some advantages. For example, if something is rotating at a constant angular speed, the **L** vector will be constant in magnitude and direction — which makes more sense than having the **L** vector rotating in the plane of the disk's rotation and constantly changing direction.

Because **L** is a 3D vector, it can point in any direction, which means that it has x, y, and z components, L_x, L_y, and L_z (which aren't vectors, just magnitudes). Refer to Figure 10-1 for a look at L_z.

L is the vector product of position **R** and linear momentum **P**, so $\mathbf{L} = \mathbf{R} \times \mathbf{P}$. You can also write L_x, L_y, and L_z at any given moment in terms of operators like this, where P_x, P_y, and P_z are the *momentum operators* (which return the momentum in the x, y, and z directions) and X, Y, and Z are the *position operators* (which return the position in the x, y, and z directions):

» $L_x = YP_z - ZP_y$

» $L_y = ZP_x - XP_z$

» $L_z = XP_y - YP_x$

You can write the momentum operators P_x, P_y, and P_z as

$$P_x = -i\hbar \frac{\partial}{\partial x}$$

$$P_y = -i\hbar \frac{\partial}{\partial y}$$

$$P_z = -i\hbar \frac{\partial}{\partial z}$$

In the same way, you can represent the position operators by their equivalent coordinates, that is,

» $X = x$

» $Y = y$

» $Z = z$

Then if you substitute these operator representations into the equations for L_x, L_y, and L_z, you get

$$L_x = -i\hbar \left(y \frac{\partial}{\partial z} - z \frac{\partial}{\partial y} \right)$$

$$L_y = -i\hbar \left(z \frac{\partial}{\partial x} - x \frac{\partial}{\partial z} \right)$$

$$L_z = -i\hbar \left(x \frac{\partial}{\partial y} - y \frac{\partial}{\partial x} \right)$$

Finding Commutators of L_x, L_y, and L_z

First, examine L_x, L_y, and L_z by taking a look at how they *commute* (that is, whether they can be measured at the same time; see Chapter 7 for an introduction to this concept). If they commute (for example, if $[L_x, L_y] = 0$), then you can measure any two of them (L_x and L_y, for example) exactly. If not, then they're subject to the uncertainty relation, and you can't measure them simultaneously exactly.

Okay, so what's the commutator of L_x and L_y? Using $L_x = YP_z - ZP_y$ and $L_y = ZP_x - XP_z$, you can write the following equation:

$$[L_x, L_y] = [YP_z - ZP_y, ZP_x - XP_z]$$

And you can also write this equation as

$$[L_x, L_y] = [YP_z, ZP_x] - [YP_z, XP_z] - [ZP_y, ZP_x] + [ZP_y, XP_z]$$
$$= Y[P_z, Z]P_x + X[Z, P_{yz}]P_y$$
$$= i\hbar(XP_y - YP_x)$$

But $XP_y - YP_x = L_z$, so $[L_x, L_y] = i\hbar L_z$. So, L_x and L_y don't commute, which means that you can't measure them both simultaneously with complete precision. You can also show that $[L_y, L_z] = i\hbar L_x$ and $[L_z, L_x] = i\hbar L_y$.

Because none of the components of angular momentum commute with each other, you can't measure any two simultaneously with complete precision. Rats.

The fact that the components of angular momentum don't commute also means that the L_x, L_y, and L_z operators can't share the same eigenstates. So, what can you do? How can you find an operator that shares eigenstates with the various components of L so that you can write the eigenstates as $|\ell, m\rangle$?

The usual trick here is that the square of the angular momentum, L^2, is a scalar, not a vector, so it'll commute with the L_x, L_y, and L_z operators, no problem:

» $[L^2, L_x] = 0$

» $[L^2, L_y] = 0$

» $[L^2, L_z] = 0$

Okay, cool, you're making progress. Because L_x, L_y, and L_z don't commute, you can't create an eigenstate that lists quantum numbers for any two of them. But because L^2 commutes with them, you can construct eigenstates that have eigenvalues for L^2 and any *one* of L_x, L_y, and L_z. By convention, the direction that's usually chosen is L_z.

Creating the Angular Momentum Eigenstates

Now's the time to create the actual eigenstates, $|\ell, m\rangle$, of angular momentum states in quantum mechanics. These eigenstates give you the eigenvalues, which let you solve for the Hamiltonian to get the allowed energy levels of an object with angular momentum.

TIP

Don't make the assumption that the eigenstates are $|\ell, m\rangle$. I'm going to rewrite them as $|\alpha, \beta\rangle$, where the eigenvalue of L^2 is $L^2|\alpha, \beta\rangle = \hbar^2\alpha|\alpha, \beta\rangle$. So, the eigenvalue of L^2 is $\hbar^2\alpha$, where you have yet to solve for α. Similarly, the eigenvalue of L_z is $L_z|\alpha, \beta\rangle = \hbar\beta|\alpha, \beta\rangle$.

To proceed further, you can borrow a familiar process from Chapter 9 that introduces lowering and raising operators, applying the lowering operator to the ground state, and solving for the ground state.

These operators act on the L_z quantum number, and you can define the raising and lowering operators this way:

>> Raising: $L_+|\alpha, \beta\rangle = c|\alpha, \beta + 1\rangle$

>> Lowering: $L_-|\alpha, \beta\rangle = d|\alpha, \beta - 1\rangle$

So, the raising operator, L_+, has the effect of raising the β quantum number by 1. Similarly, the lowering operator, L_-, lowers the β quantum number by 1. Neither operator changes the α quantum number at all. You will be able to determine the values of c and d in the section "Finding the Eigenvalues of the Raising and Lowering Operators," later in this chapter.

Finding the Angular Momentum Eigenvalues

The eigenvalues of the angular momentum are the possible values the angular momentum can take, so they're worth finding. You use the minimum and maximum values of β (called β_{max} and β_{min}) to find the possible quantum states of the angular momentum. You then work through an example of a *diatomic molecule*, a molecule made up of two atoms rotating around each other. The following section shows you how to do that.

Deriving eigenstate equations with β_{max} and β_{min}

Here, you look for the eigenstate equations for quantum angular momentum. To do that, you apply the trick of finding a lower and upper bound on the angular momentum. Fortunately, since you can't have a negative total angular momentum, such bounds do exist.

1. Write a bra-ket equation with a lower bound.

The angular momentum operator is the sum of the three component operators, so it turns out that $L^2 - L_z^2 = L_x^2 + L_y^2$, which is a non-negative number, so $L^2 - L_z^2 \geq 0$. This makes sense if you think about it: The total angular momentum minus just the z component of the angular momentum cannot be less than 0.

This means that you can write this equation:

$$\langle \alpha, \beta \,|\, (L^2 - L_z^2) \,|\, \alpha, \beta \rangle \geq 0$$

2. Apply the operators.

Substituting in $L^2 |\alpha, \beta\rangle = \alpha \hbar^2 |\alpha, \beta\rangle$ and $L_z |\alpha, \beta\rangle = \beta \hbar |\alpha, \beta\rangle$ (twice), and using the fact that the eigenstates are normalized, gives you this:

$$\langle \alpha, \beta \,|\, \left(L^2 - L_z^2\right) |\alpha, \beta\rangle = \hbar^2 \left(\alpha - \beta^2\right) \geq 0$$

From the last equation, you can find that $\alpha \geq \beta^2$. So, there's a maximum possible value of β, which you can call β_{max}.

3. Apply the raising and lowering operators.

You can be clever now, because there has to be a state $|\alpha, \beta_{max}\rangle$ such that you can't raise β any higher. Thus, if you apply the raising operator, you get zero:

$$L_+ |\alpha, \beta_{max}\rangle = 0$$

Applying the lowering operator to this equation also gives you zero:

$$L_- L_+ |\alpha, \beta_{max}\rangle = 0$$

4. Translate to angular momentum operators.

Here you have to use another relationship, which comes from applying the lowering operator to the raising operator, $L_- L_+ = L^2 - L_z^2 - \hbar L_z$. (You can confirm this yourself in *Quantum Physics Workbook For Dummies* [Wiley].) Applying that relationship here means the following is true:

$$\left(L^2 - L_z^2 - \hbar L_z\right) |\alpha, \beta_{max}\rangle = 0$$

Putting in $L^2|\alpha, \beta_{max}\rangle = \alpha \hbar^2$ and $L_z|\alpha, \beta_{max}\rangle = \beta_{max}\hbar|\alpha, \beta_{max}\rangle$ gives you this:

$$\left(\alpha - \beta_{max}{}^2 - \beta_{max}\right)\hbar^2 = 0$$

$$\alpha = \beta_{max}\left(\beta_{max} + 1\right)$$

You've now defined α in terms of β_{max}. At this point, it's typical to rename β_{max} as ℓ and β as m (which, as a reminder from the beginning of the chapter, is the quantum number for the z component of angular momentum, and does not represent mass), so $|\alpha, \beta\rangle$ becomes $|\ell, m\rangle$ and

- $L^2|\ell, m\rangle = \ell(\ell + 1)\hbar^2|\ell, m\rangle$

- $L_z|\ell, m\rangle = m\hbar|\ell, m\rangle$

You can say even more, applying much the same reasoning as in the previous steps.

From the $\alpha \geq \beta^2$ in Step 2, you can tell that in addition to a β_{max}, there must also be a β_{min} such that when you apply the lowering operator, L_-, you get zero, because you can't go any lower than β_{min}. Steps 3 and 4, in this case, get you to this equation:

$$\left(\alpha - \beta_{min}{}^2 + \beta_{min}\right)\hbar^2 = 0$$

$$\alpha - \beta_{min}{}^2 + \beta_{min} = 0$$

$$\alpha = \beta_{min}{}^2 - \beta_{min}$$

$$\alpha = \beta_{min}\left(\beta_{min} - 1\right)$$

And comparing this equation to $\alpha = \beta_{max}(\beta_{max} + 1)$ gives you

$$\beta max = -\beta min$$

Further investigating the quantum numbers

After applying a lot of math (see the preceding step list), you can refocus on what the results mean. The second component of the eigenstate (the part that maximizes and minimizes the β quantum number) is the Z component — the direction — of the momentum. So, the maximum quantum number of the momentum's direction is in the opposite direction from the minimum quantum number.

Because of how you've defined the lowering operator L_-, you could take β_{max} and use L_- some discrete number of times, n, and eventually you'd reach β_{min}.

$$\beta_{max} = \beta_{min} + n$$

Since you know they're symmetric (because $\beta_{max} = -\beta_{min}$), this means that

$$\beta_{max} = \frac{n}{2}$$

Therefore, β_{max} can be either an integer (if n is even) or a half-integer (if n is odd).

REMEMBER

Because $\ell = \beta_{max},$ $m = \beta$, and n is a positive number, you can find that $-\ell \le m \le \ell$. So now you have a lot of information about this quantum angular momentum situation:

>> The eigenstates are $|\ell, m\rangle$.

>> The quantum number of the total angular momentum is ℓ.

>> The quantum number of the angular momentum along the z-axis is m.

>> $L^2|\ell, m\rangle = \hbar^2\ell(\ell+1)|\ell, m\rangle$ where $\ell = 0, \frac{1}{2}, 1, \frac{3}{2},\ldots$

>> $L_z|\ell, m\rangle = \hbar m|\ell, m\rangle$, where $m = -\ell, -(\ell-1),\ldots, \ell-1, \ell$.

>> $-\ell \le m \le \ell$.

For each ℓ, you have $2\ell + 1$ values of m. For example, if $\ell = 2$, then m can equal -2, $-1, 0, 1,$ or 2. If $\ell = \frac{5}{2}$, then m can equal $-\frac{5}{2}, -\frac{3}{2}, -\frac{1}{2}, \frac{1}{2}, \frac{3}{2},$ and $\frac{5}{2}$.

Looking at the results intuitively

REMEMBER

In addition to the evidence of the underlying mathematics, consider this: If the angular momentum is higher in general, then you'd expect the amount of angular momentum in the Z direction to also be higher. Intuitively, this feels like the sort of result you'd expect. Quantum physics relies so much on obtuse mathematical formalism that it's nice when, every once in a while, the results actually seem intuitively correct!

You can see a representative L and L_z in Figure 10-2. L is the total angular momentum and L_z is the projection of that total angular momentum on the z-axis.

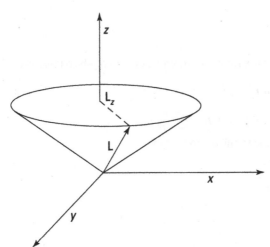

FIGURE 10-2:
L and L_z.

Getting the rotational energy of a diatomic molecule

In this section, you look at an example that involves finding the rotational energy spectrum of a diatomic molecule. Figure 10-3 shows the setup: A rotating diatomic molecule is composed of two atoms with masses m_1 and m_2 (not to be confused with the quantum number m in the last section, even though it is, unfortunately, confusing). The first atom rotates at $r = r_1$, and the second atom rotates at $r = r_2$. What's the molecule's rotational energy?

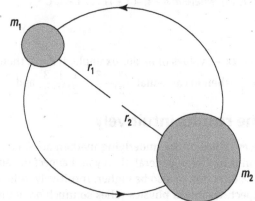

FIGURE 10-3:
A rotating diatomic molecule.

1. **Establish the Hamiltonian and related variables.**

   Start by making sure you have the variables you need. The Hamiltonian for angular momentum (as described in the section "Setting Up the Hamiltonian," earlier in the chapter) is

   $$H = \frac{L^2}{2I}$$

   The term I is the rotational moment of inertia, which looks like

   $$I = m_1 r_1^2 + m_2 r_2^2 = \mu r^2$$

   where $r = |r_1 - r_2|$ and $\mu = \dfrac{m_1 m_2}{m_1 + m_2}$.

   Therefore, the Hamiltonian becomes

   $$H = \frac{L^2}{2I} = \frac{L^2}{2\mu r^2}$$

2. **Apply the Hamiltonian to eigenstate vectors to solve for E.**

So, applying the Hamiltonian to the eigenstates, $|\ell, m\rangle$, gives you the following:

$$H|\ell, m\rangle = \frac{L^2}{2\mu r^2}|\ell, m\rangle$$

From the section "Further investigating the quantum numbers," you know that $L^2|\ell, m\rangle = \ell(\ell+1)\hbar^2|\ell, m\rangle$, so this equation becomes

$$H|\ell, m\rangle = \frac{L^2}{2\mu r^2}|\ell, m\rangle = \frac{\ell(\ell+1)\hbar^2}{2\mu r^2}|\ell, m\rangle$$

And because $H|\ell, m\rangle = E|\ell, m\rangle$, you can see that

$$E = \frac{\ell(\ell+1)\hbar^2}{2\mu r^2}$$

REMEMBER

And that's the energy as a function of ℓ, the angular momentum quantum number, for a diatomic molecule. It's worth remembering that the masses of the two atoms do come into play, since they determine the value of μ.

Finding the Eigenvalues of the Raising and Lowering Operators

This section looks at finding the eigenvalues of the raising and lowering angular momentum operators, which I introduce in the section "Creating the Angular Momentum Eigenstates," earlier in this chapter. As you might expect, these operators raise and lower a state's z component of angular momentum.

Start by taking a look at L_+, and planning to solve for c:

$$L_+|\ell, m\rangle = c|\ell, m+1\rangle$$

So, multiplying this new state by its transpose should give you c^2:

$$(L_+|\ell, m\rangle)^\dagger L_+|\ell, m\rangle = (c\langle\ell, m+1|)(c|\ell, m+1\rangle)$$
$$(L_+|\ell, m\rangle)^\dagger L_+|\ell, m\rangle = c^2\langle\ell, m+1|\ell, m+1\rangle$$
$$(L_+|\ell, m\rangle)^\dagger L_+|\ell, m\rangle = c^2$$

You can also write the transpose in terms of a raising and lowering operator, like this:

$$(L_+|\ell, m\rangle)^{\dagger}\, L_+|\ell, m\rangle = \langle l, m\,|\,L_+L_-\,|\,l, m\rangle = c^2$$

What do you do about $L_+\, L_-$? To help, I give you another useful relationship (which you can prove for yourself in *Quantum Physics Workbook For Dummies*): $L_+L_- = L^2 - L_z^2 + \hbar L_z$. Substitute this relationship, and your equation becomes the following, which you can solve for c:

$$\langle \ell, m|\left(L^2 - L_z^2 + \hbar L_z\right)|\ell, m\rangle = c^2$$

$$c = \left(\langle \ell, m|\left(L^2 - L_z^2 + \hbar L_z\right)|\ell, m\rangle\right)^{1/2}$$

$$c = \hbar[\ell(\ell+1) - m(m+1)]^{1/2}$$

And that's the eigenvalue of L_+. Performing a similar process for L_- gets you that eigenvalue, d. This results in the following two relations:

$$L_+|\ell, m\rangle = \hbar\sqrt{\ell(\ell+1) - m(m+1)}\,|\ell, m+1\rangle$$

$$L_-|\ell, m\rangle = \hbar\sqrt{\ell(\ell+1) - m(m-1)}\,|\ell, m-1\rangle$$

Interpreting Angular Momentum with Matrices

In this section, you get to take a look at the matrix representation of angular momentum on a quantum level. Don't worry, I stick with small quantum numbers, so the matrices don't get too big. The matrix method was the first system of quantum mechanics. Physicists recognized it as extremely unwieldy right from the start and ultimately replaced the matrix method with the wave mechanics and Dirac notation for a reason: simpler calculations!

Consider a system with the total angular momentum quantum number $\ell = 1$, which means that m can take the values -1, 0, and 1. So, you can represent the three possible angular momentum states like this:

$$|1, -1\rangle = \begin{bmatrix} 0 \\ 0 \\ 1 \end{bmatrix} \qquad |1, 0\rangle = \begin{bmatrix} 0 \\ 1 \\ 0 \end{bmatrix} \qquad |1, 1\rangle = \begin{bmatrix} 1 \\ 0 \\ 0 \end{bmatrix}$$

Okay, so what are the operators you've seen in this chapter in matrix representation? For example, what is L^2? You can write L^2 this way in matrix form:

$$L^2 = \begin{bmatrix} \langle 1,1|L^2|1,1\rangle & \langle 1,1|L^2|1,0\rangle & \langle 1,1|L^2|1,-1\rangle \\ \langle 1,0|L^2|1,1\rangle & \langle 1,0|L^2|1,0\rangle & \langle 1,0|L^2|1,-1\rangle \\ \langle 1,-1|L^2|1,1\rangle & \langle 1,-1|L^2|1,0\rangle & \langle 1,-1|L^2|1,-1\rangle \end{bmatrix}$$

This matrix might look a bit terrifying, but take a deep breath and work the problem in pieces. Just follow these steps:

1. Identify any elements that equal zero.

All of the vectors except for the diagonal are combining with an *orthogonal vector*, which means they are equal to zero. So those are easy enough. Only the elements along the diagonal have non-zero values to worry about.

2. Combine the vectors along the diagonal. (Ignore the L^2 operator for now.)

The vectors are combining with their transpose, and they're normalized, so the result is

$$\langle 1,1|1,1\rangle = 1 \qquad \langle 1,0|1,0\rangle = 1 \qquad \langle 1,-1|1,-1\rangle = 1$$

3. Apply the L^2 operator.

The momentum operator, L^2, results in $L^2|\ell, m\rangle = \hbar^2 \ell(\ell+1)|\ell, m\rangle$ (which you calculated in the section "Deriving eigenstate equations with β_{max} and β_{min}," earlier in the chapter).

Since $\ell = 1$, you know that $\hbar^2 \ell(\ell+1) = 2\hbar^2$. So, you can write the matrix for L^2 as

$$L^2 = \begin{bmatrix} 2\hbar^2 & 0 & 0 \\ 0 & 2\hbar^2 & 0 \\ 0 & 0 & 2\hbar^2 \end{bmatrix} = 2\hbar^2 \begin{bmatrix} 1 & 0 & 0 \\ 0 & 1 & 0 \\ 0 & 0 & 1 \end{bmatrix}$$

So, in matrix form, the equation $L^2|1, 1\rangle = 2\hbar^2|1, 1\rangle$ becomes

$$2\hbar^2 \begin{bmatrix} 1 & 0 & 0 \\ 0 & 1 & 0 \\ 0 & 0 & 1 \end{bmatrix} \begin{bmatrix} 1 \\ 0 \\ 0 \end{bmatrix} = 2\hbar^2 \begin{bmatrix} 1 \\ 0 \\ 0 \end{bmatrix}$$

You can apply similar reasoning (see the next two sections) with other operators discussed earlier in this chapter.

Raising and lowering operators

How about the L_+ operator? As you probably know (from the section "Finding the Eigenvalues of the Raising and Lowering Operators," earlier in the chapter), $L_+|\ell, m\rangle = \hbar\sqrt{\ell(\ell+1) - m(m+1)}|\ell, m+1\rangle$. In this example, $\ell = 1$ and $m = +1$, 0, and -1. So, you have the following:

$$L_+|1, 1\rangle = 0 \qquad L_+|1, 0\rangle = \sqrt{2}\hbar|1, 1\rangle \qquad L_+|1, -1\rangle = \sqrt{2}\hbar|1, 0\rangle$$

In matrix form, then, consider that the L_+ operator will turn the first vector into 0, but will increase the m value in each of the other columns by 1, and multiply by the scalar. This looks like the following:

$$L_+ = \sqrt{2}\hbar \begin{bmatrix} 0 & 1 & 0 \\ 0 & 0 & 1 \\ 0 & 0 & 0 \end{bmatrix}$$

Okay, what about L_-? Applying the same reasoning for that operator results in this matrix form:

$$L_- = \sqrt{2}\hbar \begin{bmatrix} 0 & 0 & 0 \\ 1 & 0 & 0 \\ 0 & 1 & 0 \end{bmatrix}$$

You can test out both of those operator matrices with the different vectors to confirm that they work out with the same results as applying the operators in the Dirac notation.

Moving on to other L operators

With the matrices for L^2, L_+, and L_- established, you can move forward to thinking about the matrix representation for L_z. This is simple because

>> $\hbar|1, 1\rangle = L_z|1, 1\rangle$

>> $0 = L_z|1, 0\rangle$

>> $-\hbar|1, 1\rangle = L_z|1, -1\rangle$

So, the first column of L_z yields the scalar $\hbar$, the middle column yields 0, and the last column yields $-\hbar$. The resulting matrix is

$$L_z = \hbar \begin{bmatrix} 1 & 0 & 0 \\ 0 & 0 & 0 \\ 0 & 0 & -1 \end{bmatrix}$$

And you should, of course, test this one out to confirm that it works with the various vectors, as well.

Similar logic works to find the L_x, L_y, and $[L_x, L_y]$ operators. That's not as hard as you may think, using the following as your starting points, and then applying matrix addition:

$$L_x = \frac{1}{2}(L_+ + L_-)$$

$$L_y = \frac{i}{2}(L_- - L_+)$$

$$[L_x, L_y] = L_x L_y - L_y L_x$$

Rounding It Out: Switching to the Spherical Coordinate System

This chapter's first several sections deal with angular momentum in *Dirac notation* (a language designed to fit the precise needs of expressing states in quantum mechanics; see its introduction in Chapter 7). The charm of this notation, with its bras and kets, is that it doesn't limit you to any specific system of representation. You can focus on the bras and kets directly or translate them into matrices if you prefer to work that way.

So, you have the general eigenstates, which you can work with easily enough (relatively speaking) in Dirac notation, but what if you want to dive more deeply into the particulars? What if, instead of just the general eigenstates, you want to find the actual eigenfunctions of L_z and L^2, the actual functions that you can use with the angular momentum operators like L^2 and L_z? Well, if you plan to dive into those particulars, you need a little change in perspective.

Laying the (spherical) groundwork

TIP

Even in classical physics, when talking about angular momentum, it makes sense to translate your thinking into spherical coordinates (or polar coordinates in two dimensions) instead of a traditional Cartesian system of rectangular coordinates. Chapter 13 is entirely devoted to spherical coordinates, but I touch on them here for the sake of simplicity. The math for angular momentum quickly becomes a nightmare in a rectangular system, but it is greatly simplified in these other coordinate systems. Figure 10-4 shows the spherical coordinate system.

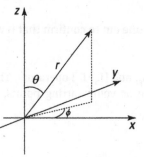

FIGURE 10-4:
The spherical
coordinate
system.

In the rectangular (Cartesian) coordinate system, you use x, y, and z to orient yourself. In the spherical coordinate system, you also use three quantities: r, θ, and ϕ, as Figure 10-4 shows. You can translate between the spherical coordinate system and the rectangular one. The r vector is the vector to the particle that has angular momentum, θ is the angle of r from the z-axis, and ϕ is the angle of r from the x-axis. The following equations show how you can write the three rectangular coordinates in terms of the spherical coordinates.

» $x = r \sin\theta \cos\phi$

» $y = r \sin\theta \sin\phi$

» $z = r \cos\phi$

Consider the equations for angular momentum:

$$L_x = YP_z - ZP_y = -i\hbar\left(y\frac{\partial}{\partial z} - z\frac{\partial}{\partial y} \right)$$

$$L_y = ZP_x - XP_z = -i\hbar\left(z\frac{\partial}{\partial x} - x\frac{\partial}{\partial z} \right)$$

$$L_z = XP_y - YP_x = -i\hbar\left(x\frac{\partial}{\partial y} - y\frac{\partial}{\partial x} \right)$$

When you take the angular momentum equations with the spherical-coordinate-system conversion equations, you can derive the following:

» $L_z = -i\hbar\dfrac{\partial}{\partial \phi}$

» $L^2 = -\hbar^2\left(\dfrac{1}{\sin\theta}\dfrac{\partial}{\partial \theta}\left(\sin\theta\dfrac{\partial}{\partial \theta} \right) + \dfrac{1}{\sin^2\theta}\dfrac{\partial^2}{\partial \phi^2} \right)$

» $L_+ = L_x + iL_y = \hbar e^{i\phi}\left(\dfrac{\partial}{\partial \theta} + \dfrac{i\cos\theta}{\sin\theta}\dfrac{\partial}{\partial \phi} \right)$

» $L_- = L_x - iL_y = \hbar e^{-i\phi}\left(-\dfrac{\partial}{\partial \theta} + \dfrac{i\cos\theta}{\sin\theta}\dfrac{\partial}{\partial \phi} \right)$

These equations look pretty involved. Take note of this one extremely significant feature: The variable r is completely absent. They depend only on θ and ϕ, which means their eigenstates depend only on θ and ϕ, not on r. So, the eigenfunctions of the operators in the preceding list can be denoted like this:

$$\langle \theta, \phi | l, m \rangle$$

Traditionally, you give the name $Y_{\ell m}(\theta, \phi)$ to the eigenfunctions of angular momentum in spherical coordinates, so you have the following:

$$Y_{\ell m}(\theta, \phi) = \langle \theta, \phi | \ell, m \rangle$$

REMEMBER

Keep in mind that the constraints on quantum numbers ℓ and m will show up in the index for $Y_{\ell m}$. The quantization in quantum physics means that these numbers are always taking on discrete integer values, because that's how quantum numbers work.

In other words, if you use an example (from the section "Raising and lowering operators," earlier in the chapter) where $\ell = 1$ and $m = +1, 0,$ and -1 (being a little bit more explicit about the positive m value here), then you end up with the following $Y_{\ell m}$ functions to consider: $Y_{1,+1}$, $Y_{1,0}$, and $Y_{1,-1}$. For a glimpse ahead, you see that some of the symmetries in these equations mean that you can combine the first and last functions into a single function notation: $Y_{1,\pm 1}$. Simplifying with this combination is another great reason to be using spherical coordinates!

All right, now that you have a sense of what to look for, you can work on finding the actual form of $Y_{\ell m}(\theta, \phi)$. When using the L^2 and L_z operators on angular momentum eigenstates, earlier work showed that these relationships will hold:

$$L^2 |\ell, m\rangle = \ell(\ell+1)\hbar^2 |\ell, m\rangle$$
$$L_z |\ell, m\rangle = m\hbar |\ell, m\rangle$$

So, the following must be true:

» $L^2 Y_{\ell m}(\theta, \phi) = \ell(\ell+1)\hbar^2 Y_{\ell m}(\theta, \phi)$

» $L_z Y_{\ell m}(\theta, \phi) = m\hbar Y_{\ell m}(\theta, \phi)$

In fact, you can go further. Remember that L_z is the Z component of the angular momentum, which means that L_z depends only on θ (the angle relative to the z-axis) and not at all on ϕ (the angle relative to the x-axis). This suggests that you can split $Y_{\ell m}(\theta, \phi)$ into a part that depends on θ and a part that depends on ϕ. Splitting $Y_{\ell m}(\theta, \phi)$ into parts looks like this:

$$Y_{\ell m}(\theta, \phi) = \Theta_{\ell m}(\theta)\Phi_m(\phi)$$

This splitting of the eigenfunctions is what makes working with spherical coordinates so helpful. Splitting isn't possible when you work in rectangular coordinates. Even before you see how to proceed with the more detailed work, you must clearly understand that you can exploit splitting the eigenfunction in a positive way to reach your solutions.

The eigenfunctions of L_z in spherical coordinates

Start by finding the eigenfunctions of L_z in spherical coordinates. As a reminder (see the previous section), that L_z operator looks like this:

$$L_z = -i\hbar\frac{\partial}{\partial\phi}$$

Doesn't look so bad, does it? Begin by applying the operator to $Y_{\ell m}(\theta, \phi)$, and then breaking it apart to use the different parts of the function. (*Note:* Here is where you see the big benefit of splitting up those spherical functions.) Because the L_z operator is a derivative with respect to ϕ, the part of the function that is dependent solely on θ comes outside of the derivative.

$$L_z Y_{lm}(\theta, \phi) = L_z \Theta_{lm}(\theta)\Phi_m(\phi)$$

$$L_z Y_{lm}(\theta, \phi) = -i\hbar\frac{\partial}{\partial\phi}\Theta_{lm}(\theta)\Phi_m(\phi)$$

$$L_z Y_{lm}(\theta, \phi) = -i\hbar\Theta_{lm}(\theta)\frac{\partial\Phi_m}{\partial\phi}(\phi)$$

But this isn't all that you know about L_z. From the previous section, you saw that $L_z Y_{\ell m}(\theta, \phi) = m\hbar Y_{\ell m}(\theta, \phi)$ and $Y_{\ell m}(\theta, \phi) = \Theta_{\ell m}(\theta)\Phi_m(\phi)$.

$$-i\hbar\Theta_{\ell m}(\theta)\frac{\partial\Phi_m}{\partial\phi}(\phi) = m\hbar\Theta_{\ell m}(\theta)\Phi_m(\phi)$$

$$-i\frac{\partial\Phi_m}{\partial\phi}(\phi) = m\Phi_m(\phi)$$

This looks easy to solve (okay, maybe it doesn't really look easy, but the solution can be found), and the solution is just

$$\Phi_m(\phi) = Ce^{im\phi}$$

where C is a constant of integration. You can determine C by insisting that $\Phi_m(\phi)$ be normalized — that is, that the following holds true:

$$\int_0^{2\pi} \Phi_m^*(\phi)\Phi_m(\phi)d\phi = 1$$

198 PART 3 By the Numbers: Basic Quantum Physics Math

which gives you $C = \dfrac{1}{\sqrt{2\pi}}$.

So, $\Phi_m(\phi)$ is equal to this:

$$\Phi_m(\phi) = \frac{e^{im\phi}}{\sqrt{2\pi}}$$

You're making progress — you've determined the form of $\Phi_m(\phi)$, so $Y_{\ell m}(\theta, \phi) = \Theta_{\ell m}(\theta)\,\Phi_m(\phi)$, which equals

$$Y_{\ell m}(\theta, \phi) = \Theta_{\ell m}(\theta)\Phi_m(\phi) = \Theta_{\ell m}(\theta)\frac{e^{im\phi}}{\sqrt{2\pi}}$$

That's great — you're halfway there, but you still have to determine the form of $\Theta_{\ell m}(\theta)$, the eigenfunction of L^2. That happens in the next section.

The eigenfunctions of L² in spherical coordinates

You can build on the work covered in the previous section and tackle the eigen-function of L^2, $\Theta_{\ell m}(\theta)$. You already know (see the section "Laying the [spherical] groundwork") that in spherical coordinates, the L^2 operator looks like this:

$$L^2 = -\hbar^2\left(\frac{1}{\sin\theta}\frac{\partial}{\partial\theta}\left(\sin\theta\frac{\partial}{\partial\theta}\right) + \frac{1}{\sin^2\theta}\frac{\partial^2}{\partial\phi^2}\right)$$

That's quite an operator. You also know that

$$Y_{\ell m}(\theta, \phi) = \Theta_{\ell m}(\theta)\frac{e^{im\phi}}{\sqrt{2\pi}}$$

You then go through these steps to find the eigenfunctions:

1. **Apply the L² operator.**

   So, applying the L^2 operator to $Y_{\ell m}(\theta, \phi)$ gives you the following:

   $$L^2 Y_{\ell m}(\theta, \phi) = -\hbar^2\left[\frac{1}{\sin\theta}\frac{\partial}{\partial\theta}\left(\sin\theta\frac{\partial}{\partial\theta}\right) + \frac{1}{\sin^2\theta}\frac{\partial^2}{\partial\phi^2}\right]\left(\Theta_{\ell m}(\theta)\frac{e^{im\phi}}{\sqrt{2\pi}}\right)$$

2. Rewrite $Y_{\ell m}$ and simplify.

And because $L^2Y_{\ell m}(\theta, \phi) = \ell(\ell + 1)\hbar^2 Y_{\ell m}(\theta, \phi) = \ell(\ell + 1)\hbar^2\Theta_{\ell m}(\theta)\Phi_m(\phi)$, this equation becomes

$$-\hbar^2\left[\frac{1}{\sin\theta}\frac{\partial}{\partial\theta}\left(\sin\theta\frac{\partial}{\partial\theta}\right) + \frac{1}{\sin^2\theta}\frac{\partial^2}{\partial\phi^2}\right]\left(\Theta_{\ell m}(\theta)\frac{e^{im\phi}}{\sqrt{2\pi}}\right)$$

$$= \ell(\ell + 1)\hbar^2\Theta_{\ell m}(\theta)\frac{e^{im\phi}}{\sqrt{2\pi}}$$

What have you gotten into? Cancelling terms and subtracting the right-hand side from the left finally gives you this differential equation (which you can then simplify):

$$\left[\frac{1}{\sin\theta}\frac{\partial}{\partial\theta}\left(\sin\theta\frac{\partial}{\partial\theta}\right) + \frac{1}{\sin^2\theta}\frac{\partial^2}{\partial\phi^2}\right]\left(\Theta_{\ell m}(\theta)e^{im\phi}\right) + \ell(\ell + 1)\Theta_{\ell m}(\theta)e^{im\phi} = 0$$

$$\left[\frac{1}{\sin\theta}\frac{\partial}{\partial\theta}\left(\sin\theta\frac{\partial}{\partial\theta}\right) - \frac{m^2}{\sin^2\theta} + \ell(\ell + 1)\right]\Theta_{\ell m}(\theta) = 0$$

3. Solve the differential equation.

I forgive you for hoping that you don't have to do all the heavy lifting of solving this equation. And you're in luck! This equation is a *Legendre differential equation*, and the solutions are well known. In general, the solutions take this form:

$$\Theta_{\ell m}(\theta) = C_{\ell m}P_{\ell m}(\cos\theta)$$

where $P_{\ell m}(\cos\theta)$ is the *Legendre function*.

LEGENDRE DIFFERENTIAL EQUATIONS, FUNCTIONS, AND POLYNOMIALS

The Legendre differential equations are a well-known form of differential equation, and mathematicians have solved this body of differential equations for you. Yay for the mathematicians!

You can start by separating out the m dependence, which works this way with the Legendre functions:

$$P_{\ell m}(x) = \left(1 - x^2\right)^{|m|/2}\frac{d^{|m|}}{dx^{|m|}}P_\ell(x)$$

where $P_\ell(x)$ is called a *Legendre polynomial* and is given by the Rodrigues formula:

$$P_\ell(x) = \frac{(-1)^\ell}{2^\ell \ell!} \frac{d^\ell}{dx^\ell} (1 - x^2)^\ell$$

You can use this equation to derive the first few Legendre polynomials like this:

$$P_0(x) = 1$$
$$P_1(x) = x$$
$$P_2(x) = \frac{1}{2}(3x^2 - 1)$$

$$P_3(x) = \frac{1}{2}(5x^2 - 3x)$$
$$P_4(x) = \frac{1}{8}(35x^4 - 30x^2 + 3)$$
$$P_5(x) = \frac{1}{8}(63x^5 - 70x^3 + 15x)$$

and so on. That's what the first few $P_\ell(x)$ polynomials look like. So, what do the associated Legendre functions, $P_{\ell m}(x)$ look like? You can also calculate them. You can start off with $P_{\ell 0}(x)$, where $m = 0$. Those are easy because $P_{\ell 0}(x) = P_\ell(x)$, so

- $P_{10}(x) = x$
- $P_{20}(x) = \frac{1}{2}(3x^2 - 1)$
- $P_{30}(x) = \frac{1}{2}(5x^3 - 3x)$

So, for the $P_{\ell m}(x)$ where $m \neq 0$, you do have to go back up to the full Legendre function (not just the polynomial) and find them, but fortunately, you're most of the way there once you've found the associated $P_\ell(x)$. Here are a few examples of some of the handier Legendre polynomials that you might need:

- $P_{11}(x) = \sqrt{1 - x^2}$
- $P_{21}(x) = 3x\sqrt{1 - x^2}$
- $P_{22}(x) = 3(1 - x^2)$
- $P_{31}(x) = \frac{3}{2}(5x^2 - 1)\sqrt{1 - x^2}$
- $P_{32}(x) = 15x(1 - x^2)$
- $P_{33}(x) = 15x\sqrt{(1 - x^2)^3}$

These equations give you an overview of what the $P_{\ell m}$ functions look like. You can dive more into their analysis in resources on differential equations, such as

- *Differential Equations For Dummies* (Wiley)
- https://mathworld.wolfram.com/LegendrePolynomial.html
- Intro to Legendre Polynomials video: https://youtu.be/djlUmwbPw20?si=uuPbC8Dk1QFq1Vbo

4. Normalize to find the constants.

And now you know what the $P_{\ell m}$ functions look like (at least if you read the sidebar), but what do $C_{\ell m}$, the constants, look like? As soon as you have those, you'll have the complete angular momentum eigenfunctions, $Y_{\ell m}(\theta, \phi)$, because $Y_{\ell m}(\theta, \phi) = \Theta_{\ell m}(\theta)\Phi_m(\phi)$.

You can go about calculating the constants $C_{\ell m}$ the way you always calculate such constants of integration in quantum physics — you normalize the eigenfunctions to 1. For $Y_{\ell m}(\theta, \phi) = \Theta_{\ell m}(\theta)\Phi_m(\phi)$, that looks like this:

$$\int_0^{2\pi}\int_0^{\pi} Y_{\ell m}^*(\theta, \phi)Y_{\ell m}(\theta, \phi)\sin\theta\, d\theta d\phi = 1$$

Substitute the following three quantities in this equation:

- $Y_{\ell m}(\theta, \phi) = \Theta_{\ell m}(\theta)\Phi_m(\phi)$
- $\Phi_m(\phi) = \dfrac{e^{im\phi}}{\sqrt{2\pi}}$
- $\Theta_{\ell m}(\theta) = C_{\ell m}P_{\ell m}(\cos\theta)$

You get the following:

$$\frac{|C_{\ell m}|^2}{2\pi}\int_0^{2\pi} d\phi \int_0^{\pi}|P_{\ell m}(\cos\theta)|^2 \sin\theta\, d\theta = 1$$

The integral over ϕ gives 2π, so this becomes

$$|C_{\ell m}|^2\int_0^{\pi}|P_{\ell m}(\cos\theta)|^2 \sin\theta\, d\theta = 1$$

You can evaluate the integral to this, and then solve for $C_{\ell m}$:

$$|C_{\ell m}|^2 \frac{2}{2\ell+1}\frac{(\ell+|m|)!}{(\ell-|m|)!} = 1$$

$$C_{\ell m} = (-1)^{|m|}\sqrt{\frac{(2\ell+1)(\ell-|m|)!}{2(\ell+|m|)!}}$$

Which means that

$$\Theta_{\ell m}(\theta) = (-1)^{|m|}\sqrt{\frac{(2\ell+1)(\ell-|m|)!}{2(\ell+|m|)!}}P_{\ell m}(\cos\theta)$$

So, $Y_{\ell m}(\theta, \phi) = \Theta_{\ell m}(\theta)\Phi_m(\phi)$, which is the angular momentum eigenfunction in spherical coordinates, is

$$Y_{\ell m}(\theta,\phi) = (-1)^{|m|}\sqrt{\frac{(2\ell+1)(\ell-|m|)!}{4\pi(\ell+|m|)!}}P_{\ell m}(\cos\theta)e^{im\phi}$$

The functions given by this equation are called the *normalized spherical harmonics*. Here are what the first few normalized spherical harmonics look like:

» $Y_{00}(\theta, \phi) = \dfrac{1}{\sqrt{4\pi}}$

» $Y_{10}(\theta, \phi) = \sqrt{\dfrac{3}{4\pi}} \cos\theta$

» $Y_{1\pm1}(\theta, \phi) = \mp\sqrt{\dfrac{3}{8\pi}} e^{\pm i\phi} \sin\theta$

» $Y_{20}(\theta, \phi) = \sqrt{\dfrac{5}{16\pi}} \left(3\cos^2\theta - 1\right)$

» $Y_{2\pm1}(\theta, \phi) = \mp\sqrt{\dfrac{15}{8\pi}} e^{\pm i\phi} \sin\theta \cos\theta$

» $Y_{2\pm2}(\theta, \phi) = \sqrt{\dfrac{15}{32\pi}} e^{\pm 2/\theta} \sin^2\theta$

Moving back to rectangular coordinates

In the previous section, you calculated the eigenfunctions in spherical coordinates, which offers advantages in terms of simplification. But what if you wanted those eigenfunctions in rectangular coordinates anyway?

Fortunately, you have equations that you can use to move between spherical and rectangular coordinates. If you don't recall them, and don't feel like flipping back to the beginning of the spherical coordinate section, you can find them here in a slightly different form:

» $\sin\theta \cos\phi = \dfrac{x}{r}$

» $\sin\theta \sin\phi = \dfrac{y}{r}$

» $\cos\theta = \dfrac{z}{r}$

Substituting these equations into the spherical solutions for $Y_{\ell m}$ (with, at times, some agile manipulations of trigonometric identities that I gladly leave as a side project for the reader) gives you the spherical harmonics in rectangular coordinates:

» $Y_{00}(x, y, z) = \dfrac{1}{\sqrt{4\pi}}$

» $Y_{10}(x, y, z) = \sqrt{\dfrac{3}{4\pi}} \left(\dfrac{z}{r}\right)$

» $Y_{1\pm1}(x, y, z) = \mp\sqrt{\dfrac{3}{8\pi}} \left(\dfrac{x \pm iy}{r}\right)$

$$\text{» } Y_{20}(x, y, z) = \sqrt{\frac{5}{16\pi}} \left(\frac{3z^2 - r^2}{r^2} \right)$$

$$\text{» } Y_{2\pm1}(x, y, z) = \mp \sqrt{\frac{15}{8\pi}} \left(\frac{z(x \pm iy)}{r^2} \right)$$

$$\text{» } Y_{2\pm2}(x, y, z) = \mp \sqrt{\frac{15}{32\pi}} \left(\frac{x^2 - y^2 \pm 2ixy}{r^2} \right)$$

Angular momentum can be mathematically complex — particularly in quantum mechanics — but it plays a pivotal role. The study of angular momentum led to a major discovery in quantum mechanics, the existence of a property of physical particles (spin) that had never previously been anticipated. See the next chapter for more about spin.

IN THIS CHAPTER

» **Discovering spin with the Stern-Gerlach experiment**

» **Looking at eigenstates and spin notation**

» **Understanding fermions and bosons**

» **Comparing the spin operators with angular momentum operators**

» **Working with Pauli matrices**

Chapter **11**

Getting Dizzy with Spin

P hysicists suggest that orbital angular momentum is not the only kind of angular momentum present in an atom — electrons could also have intrinsic built-in angular momentum. This kind of built-in angular momentum is called spin. Whether or not electrons actually spin will never be known — they're as close to point-like particles as you can come, without any apparent internal structure. Yet the fact remains that they have intrinsic angular momentum and act *as if* they're spinning. Confusing? Welcome to quantum physics!

And that's what this chapter is about — the intrinsic, built-in quantum mechanical spin of subatomic particles. You find out about the experiment that first made physicists claim that electrons had this property. Then you look at how spin relates to the quantum numbers, eigenstates, operators, and notation that you uncover in previous chapters (Chapters 7 through 10). Finally, you discover how spin helps to classify particles in physics, and the properties related to those classifications.

Note: This chapter uses knowledge of angular momentum, so if you want to check out Chapter 10, it might not be a bad idea. You may also want to go to Chapter 2 for a refresher on electromagnetism. In addition, the equations within this chapter (which are fewer than in other chapters) still require those old tools of quantum physics: calculus, linear algebra, and differential equations.

Investigating the Stern-Gerlach Experiment and the Case of the Missing Spot

The Stern–Gerlach experiment unexpectedly revealed the existence of spin back in 1922. Physicists Otto Stern and Walther Gerlach sent a beam of silver atoms through the poles of a magnet whose magnetic field was in the vertical direction (as shown in Figure 11-1). The behavior of the atoms (refer to the figure) showed that different atoms followed different paths (spun up or down) within the magnetic field.

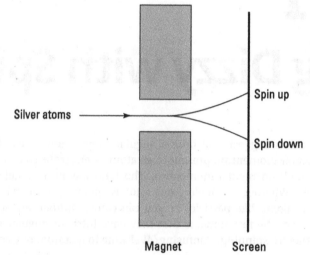

Silver atoms

Spin up

Spin down

Magnet   Screen

FIGURE 11-1:
The Stern-Gerlach experiment.

Because 46 of silver's 47 electrons are arranged in a symmetrical cloud that surrounds the atom's nucleus, any angular momentum contributions from specific electrons are cancelled out by other electrons. As a result, they contribute nothing to the total orbital angular momentum of the atom. The net angular momentum all comes down to that 47th electron because there's no paired electron to cancel out its angular momentum.

The 47th electron can be in

>> The 5s state, in which case its angular momentum is $\ell = 0$ and the z component of that angular momentum is 0.

>> The 5p state, in which case its angular momentum is $\ell = 1$, which means that the z component of its angular momentum can be –1, 0, or 1.

These possible states mean that Stern and Gerlach expected to see either one or three spots on the screen shown at right in Figure 11-1, corresponding to the different states of the z component of angular momentum.

But famously, they saw only two spots (labeled *Spin up* and *Spin down* in the figure). This disconnect — between expectation and results — puzzled the physics community for about three years. Then, in 1925, physicists Samuel A. Goudsmit and George E. Uhlenbeck suggested that electrons contained intrinsic angular momentum — and that intrinsic angular momentum is what gave them a magnetic moment that interacted with the magnetic field. After all, it was apparent that some angular momentum other than orbital angular momentum was at work in the experiment.

REMEMBER

The "other" built-in angular momentum came to be called spin. The beam of silver atoms divides in two, depending on the direction of spin of the 47th electron in the atom, so there are two possible states of spin, which came to be known as *up* and *down*.

Spin is a purely quantum mechanical effect, and no real analogue exists in classical physics. The closest you can come is to liken spin to the spin of the Earth as it goes around the Sun — that is, the Earth has both spin (because it's rotating on its axis) and orbital angular momentum (because it's revolving around the Sun). But even this picture doesn't fully explain spin in classical terms because it's conceivable that the Earth could exist without spinning. But you can't stop electrons from possessing spin, and that also goes for other subatomic particles that possess spin, such as protons.

TIP

Spin doesn't depend on spatial degrees of freedom; even if you were to have an electron at rest (which violates the uncertainty principle; see Chapter 3), it would still possess spin.

Getting Down and Dirty with Spin and Eigenstates

Spin throws a bit of a curve at you. When dealing with orbital angular momentum (see Chapter 10), you can build angular momentum operators because orbital angular momentum is the product of momentum and radius. But spin is built in; no momentum operator is involved.

So here's the crux: You cannot describe spin with a differential operator and can't find eigenfunctions for spin as you do for angular momentum. You're left with the

Dirac notation — bra and ket — way of looking at these quantum states, which isn't tied to any specific representation in spatial terms.

In Chapter 10, you also look at things in angular momentum terms, introducing the eigenstates of orbital angular momentum like this: $|\ell, m\rangle$ (where ℓ is the angular momentum quantum number and m is the quantum number of the z component of angular momentum).

You can use the same notation for spin eigenstates. As with orbital angular momentum, you can use a total spin quantum number and a quantum number that indicates the spin along the z axis. (*Note:* The z axis is always in the direction of the applied magnetic field. The spin doesn't have a set axis.)

REMEMBER

The letters given to the total spin quantum number and the z-axis component of the spin are s and m (you sometimes see them written as s and m_s). In other words, the eigenstates of spin are written as $|s, m\rangle$.

Halves and Integers: Saying Hello to Fermions and Bosons

In analogy with orbital angular momentum, you can assume that m (the z-axis component of the spin) can take the values $-s, -s + 1, \ldots, s - 1$, and s, where s is the total spin quantum number. For electrons from the silver atom (see the section "Investigating the Stern-Gerlach Experiment and the Case of the Missing Spot," earlier in the chapter), Stern and Gerlach observed two spots, so you have $2s + 1 = 2$, which means that $s = 1/2$. Therefore, m can be $+1/2$, which is an electron that is said to have an "up spin," or $-1/2$ for a "down spin." So, here are the possible eigenstates for electrons in terms of spin.

>> Spin up eigenstate: $\left|\frac{1}{2}, \frac{1}{2}\right\rangle$

>> Spin down eigenstate: $\left|\frac{1}{2}, -\frac{1}{2}\right\rangle$

So, do all subatomic particles have $s = 1/2$? Nope. Here are their options.

>> **Fermions:** In physics, particles with half-integer spin are called *fermions*. They include electrons, protons, neutrons, and so on, even quarks. For example, electrons, protons, and neutrons have spin $s = 1/2$, and there are even some particles (called delta particles) with $s = 3/2$.

>> **Bosons:** Particles with integer spin are called *bosons*. They include photons, pi mesons, and other elementary particles; even the postulated particles involved with the force of gravity, gravitons, are supposed to have integer spin. For example, pi mesons have spin $s = 0$, photons have $s = 1$, and so forth.

I cover these different types of particles more in Chapter 4, and also in the nearby sidebar, "Fermions and Bosons in Action."

So, for electrons, the spin eigenstates are $\left|\frac{1}{2}, \frac{1}{2}\right\rangle$ for spin up electrons and $\left|\frac{1}{2}, -\frac{1}{2}\right\rangle$ for spin down electrons. For photons, the eigenstates are $|1, 1\rangle$, $|1, 0\rangle$, and $|1, -1\rangle$. Therefore, the possible eigenstates depend on the particle you're working with. For the purposes of this chapter, I mostly focus on the two eigenstates, spin up and spin down, of the electron.

FERMIONS AND BOSONS IN ACTION

This chapter focuses on the quantum mechanics concept of spin, and the equations related to it, but spin is a lot more than just a number. The spin property of a particle defines certain ways that it can interact with other identical particles, which I cover in greater detail in Chapter 15.

The fermions generally represent particles that are thought of as physical objects, or particles of matter. They, too, are split into categories.

- **Leptons:** Fundamental particles of matter, the most prominent of these is the electron. It also includes the heavier muon and tau particles. All three of these particles also have neutrinos (which are also leptons) that are associated with them: the electron neutrino, the muon neutrino, and the tau neutrino.

- **Quarks:** Quarks are fundamental particles that never exist in isolation but can only be found bound together to form other particles. All particles made up of quarks are called *hadrons*. The existence of quarks is essentially deduced from the way that hadrons break apart and reform, particularly as observed in particle accelerator experiments. Since each quark has a half-integer spin, these composite hadrons can be either fermions or bosons.

- **Baryons:** When quarks bond together to form a hadron with a half-integer spin, the result is called a *baryon*. The most prominent examples are the proton and the neutron.

(continued)

(continued)

The bosons generally represent particles that are thought of as force particles, because the fundamental ones are associated with crucial forces in physics.

- **Photon:** A spin 1 particle that is known as the light particle. In addition to carrying light energy, it's also the particle that mediates the electromagnetic force between charged particles.

- **Weak boson:** A spin 1 particle that is used to mediate the weak nuclear force. There are two charged weak bosons, W^+ and W^-, and then also the neutral weak boson, Z^0.

- **Gluon:** A spin 1 particle that is used to mediate the strong nuclear force, holding quarks together to form hadrons (like protons and neutrons) and also to hold hadrons together (like keeping protons and neutrons contained in atomic nuclei).

- **Higgs boson:** A spin 0 particle predicted in 1963 and discovered at the Large Hadron Collider in 2013, which explains certain properties of the Standard Model of particle physics.

- **Meson:** When quarks bond together to form a hadron with an integer spin value, the result is called a *meson*. Examples of these are pions and kaons, although they don't come up much outside of particle accelerators and related high-energy events.

Since it hasn't been discovered (yet), I didn't include the *graviton* in the list of bosons. This particle is a purely hypothetical boson that would mediate the gravitational force — the only fundamental force in physics that isn't mediated by one of the known (and experimentally confirmed) bosons. People hoping to develop a *theory of quantum gravity* — which would explain gravity in the same quantum terms used to describe the rest of modern physics — sometimes apply the concept of a graviton as a way to approach the concept. Not having found the particle to actually exist (yet), however, is a bit of a stumbling block.

Spin Operators: Running Around with Angular Momentum

Because spin is a type of built-in angular momentum, the spin operators have a lot in common with the orbital angular momentum operators. In Chapter 10, I discuss the orbital angular momentum operators L^2 and L_z, and as you may expect, there are analogous spin operators, S^2 and S_z. However, these operators are just operators; they don't have a differential form like the orbital angular momentum operators do (which means you won't be able to find eigenfunctions for them).

Defining the spin operators

In fact, all the orbital angular momentum operators, such as L_x, L_y, and L_z, have analogues here: S_x, S_y, and S_z. The commutation relations among L_x, L_y, and L_z are the following:

» $[L_x, L_y] = i\hbar L_z$

» $[L_y, L_z] = i\hbar L_x$

» $[L_z, L_x] = i\hbar L_y$

And they work the same way for spin:

» $[S_x, S_y] = i\hbar S_z$

» $[S_y, S_z] = i\hbar S_x$

» $[S_z, S_x] = i\hbar S_y$

The L^2 operator gives you the following result when you apply it to an orbital angular momentum eigenstate:

$$L^2|\ell, m\rangle = \ell(\ell+1)\hbar^2|\ell, m\rangle$$

And just as you'd expect, the S^2 operator works in an analogous fashion:

$$S^2|s, m\rangle = s(s+1)\hbar^2|s, m\rangle$$

The L_z operator gives you this result when you apply it to an orbital angular momentum eigenstate (see Chapter 10):

$$L_z|\ell, m\rangle = m\hbar|\ell, m\rangle$$

And by analogy, the S_z operator works this way:

$$S_z|s, m\rangle = m\hbar|s, m\rangle$$

Raising and lowering spin operators

What about the raising and lowering operators, L_+ and L_-? Are there analogues for spin? In angular momentum terms, L_+ and L_- work like this, as revealed in the last chapter:

» $L_+|\ell, m\rangle = \hbar\sqrt{\ell(\ell+1) - m(m+1)}|\ell, m+1\rangle$

» $L_-|\ell, m\rangle = \hbar\sqrt{\ell(\ell+1) - m(m-1)}|\ell, m-1\rangle$

There are spin raising and lowering operators as well (S_+ and S_-), and they work exactly the same way, using the quantum number s instead of ℓ:

» $S_+|s, m\rangle = \hbar\sqrt{s(s+1) - m(m+1)}|s, m+1\rangle$

» $S_-|s, m\rangle = \hbar\sqrt{s(s+1) - m(m-1)}|s, m-1\rangle$

Working with Spin ½ and Pauli Matrices

Spin ½ particles (fermions) need a little extra attention. The eigenvalues of the S^2 operator here are

$$S^2|s, m\rangle = s(s+1)\hbar^2|s, m\rangle = \frac{3}{4}\hbar^2|s, m\rangle$$

And the eigenvalues of the S_z operator are

$$S_z|s, m\rangle = m\hbar|s, m\rangle = \pm\frac{\hbar}{2}|s, m\rangle$$

You can represent these two equations graphically, as shown in Figure 11-2, where the two spin states have different projections along the z axis.

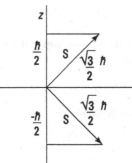

FIGURE 11-2:
Spin magnitude
and z projection.

Spin ½ matrices

Time to take a look at the spin eigenstates and operators for particles of spin ½ in terms of matrices. There are only two possible states, spin up and spin down, so this is easy.

Spin up and spin down eigenstates

In a matrix format, the entries will correspond to the positive and negative spins. When the spin is positive, the first entry will be 1 and the second entry will be 0. When the spin is negative, the numbers will be the opposite. Here are the spin eigenstates represented this way:

$$\left|\frac{1}{2}, \frac{1}{2}\right\rangle = \begin{bmatrix} 1 \\ 0 \end{bmatrix} \qquad \left|\frac{1}{2}, -\frac{1}{2}\right\rangle = \begin{bmatrix} 0 \\ 1 \end{bmatrix}$$

The matrix operator S²

Now what about spin operators like S²? The S² operator looks like this in matrix terms:

$$S^2 = \begin{bmatrix} \left\langle \frac{1}{2}, \frac{1}{2}\left|S^2\right|\frac{1}{2}, \frac{1}{2}\right\rangle & \left\langle \frac{1}{2}, \frac{1}{2}\left|S^2\right|\frac{1}{2}, -\frac{1}{2}\right\rangle \\ \left\langle \frac{1}{2}, -\frac{1}{2}\left|S^2\right|\frac{1}{2}, \frac{1}{2}\right\rangle & \left\langle \frac{1}{2}, -\frac{1}{2}\left|S^2\right|\frac{1}{2}, -\frac{1}{2}\right\rangle \end{bmatrix}$$

Multiplying the spin up and spin down vectors results in 0, so the top-right and bottom-left entries will become 0. Since the vectors are normalized, this means that the top-left and bottom-right entries will be constant. This works out to be the following:

$$S^2 = \frac{3}{4}\hbar^2 \begin{bmatrix} 1 & 0 \\ 0 & 1 \end{bmatrix}$$

The matrix operator S$_z$

Similarly, you can represent the S$_z$ operator with matrix notation:

$$S_z = \begin{bmatrix} \left\langle \frac{1}{2}, \frac{1}{2}\left|S_z\right|\frac{1}{2}, \frac{1}{2}\right\rangle & \left\langle \frac{1}{2}, \frac{1}{2}\left|S_z\right|\frac{1}{2}, -\frac{1}{2}\right\rangle \\ \left\langle \frac{1}{2}, -\frac{1}{2}\left|S_z\right|\frac{1}{2}, \frac{1}{2}\right\rangle & \left\langle \frac{1}{2}, -\frac{1}{2}\left|S_z\right|\frac{1}{2}, -\frac{1}{2}\right\rangle \end{bmatrix}$$

This works out to

$$S_z = \frac{\hbar}{2} \begin{bmatrix} 1 & 0 \\ 0 & -1 \end{bmatrix}$$

This matrix version of S_z allows you to find the z component of the spin of an eigenstate. For the spin down eigenstate, you'd apply the operator in matrix form, solve the matrix multiplication, and can then convert back into Dirac notation:

$$S_z \left| \frac{1}{2}, -\frac{1}{2} \right\rangle = \frac{\hbar}{2} \begin{bmatrix} 1 & 0 \\ 0 & -1 \end{bmatrix} \begin{bmatrix} 0 \\ 1 \end{bmatrix} = -\frac{\hbar}{2} \begin{bmatrix} 0 \\ 1 \end{bmatrix} = -\frac{\hbar}{2} \left| \frac{1}{2}, -\frac{1}{2} \right\rangle$$

Raising and lowering spin with matrices

How about the raising and lowering operators S_+ and S_-? Keep in mind what you expect to happen when these operators are applied to different electron spin eigenstate vectors:

>> Raising operator S_+ on the down spin eigenstate → Up spin eigenstate

>> Raising operator S_+ on the up spin eigenstate → Zero

>> Lowering operator S_- on the up spin eigenstate → Down spin eigenstate

>> Lowering operator S_- on the down spin eigenstate → Zero

In matrix form, these operators look like this:

$$S_+ = \hbar \begin{bmatrix} 0 & 1 \\ 0 & 0 \end{bmatrix}$$

$$S_- = \hbar \begin{bmatrix} 0 & 0 \\ 1 & 0 \end{bmatrix}$$

So, for example, you can use this matrix to check what happens when you apply the raising operator, S_+, to the spin down vector (defined in section "Spin up and spin down eigenstates"). What would you expect to happen? (*Hint:* Refer to the previous bulleted list. When you apply the raising operator to the spin down vector, you should get the up-spin eigenstate.)

Pauli matrices

Sometimes, you see the operators S_x, S_y, and S_z written in terms of Pauli matrices, σ_x, σ_y, and σ_z. Here's what the Pauli matrices look like:

$$\sigma_x = \begin{bmatrix} 0 & 1 \\ 1 & 0 \end{bmatrix}$$

$$\sigma_y = \begin{bmatrix} 0 & -i \\ i & 0 \end{bmatrix}$$

$$\sigma_z = \begin{bmatrix} 1 & 0 \\ 0 & -1 \end{bmatrix}$$

Now you can write S_x, S_y, and S_z in terms of the Pauli matrices like this:

$$S_x = \frac{\hbar}{2}\sigma_x$$

$$S_y = \frac{\hbar}{2}\sigma_y$$

$$S_z = \frac{\hbar}{2}\sigma_z$$

Whoo! And that concludes your look at spin.

4

Going 3D with Quantum Physics Calculations

Chapter **12**

Rectangular Coordinates: Solving Problems in 3D

One-dimensional problems are all very well and good, but the real world has three dimensions. Physicists are big fans of starting with easier problems in one dimension and then expanding to multiple dimensions. This chapter is all about leaving one-dimensional potentials behind and starting to look at spinless quantum mechanical particles in three dimensions. Here, you work with three dimensions in rectangular coordinates, starting with a look at the Schrödinger equation in glorious, real-life 3D. You then delve into free particles, box potentials, and harmonic oscillators, also in 3D.

Note: This chapter builds heavily on Chapters 8 and 9, taking the one-dimensional cases of energy wells and harmonic oscillators and looking at them in three dimensions. Like you do in those cases, you need to be comfortable with calculus and differential equations — as well as some trigonometry — to follow what's happening in this chapter.

Viewing the Schrödinger Equation in 3D!

I introduce the Schrödinger equation in Chapter 3, but in this chapter, I tackle it in its full, three-dimensional glory.

The *Hamiltonian* is the name for the energy operator, H. When you use H on a wave function, it returns the energy, E, of the wave function. The scalar E is the energy level, which is also the eigenvalue of the H operator. (See Chapter 7 for this calculation and how it's applied to derive the Schrödinger equation.)

Converting the Schrödinger equation into rectangular coordinates

In one dimension, the time-dependent Schrödinger equation (see Chapter 8 for information related to time dependence) looks like this:

$$\frac{-\hbar^2}{2m}\nabla^2\psi(r,t) + V(r,t)\psi(r,t) = i\hbar\frac{\partial}{\partial t}\psi(r,t)$$

This equation is already in three dimensions, because the *r* represents the position vector in three dimensions. You can convert this into rectangular coordinates, $\psi(r, t) = \psi(x, y, z, t)$. Now to do this, you also want to look at the Laplace operator (which serves to take the derivative of each variable), ∇^2, in a new way:

$$\nabla^2 = \left(\frac{\partial^2}{\partial x^2} + \frac{\partial^2}{\partial y^2} + \frac{\partial^2}{\partial z^2}\right)$$

Applying the Laplacian to the equation in rectangular coordinates, you get the 3D Schrödinger equation:

$$\frac{-\hbar^2}{2m}\left(\frac{\partial^2}{\partial x^2} + \frac{\partial^2}{\partial y^2} + \frac{\partial^2}{\partial z^2}\right)\psi(x,y,z,t) + V(r)\psi(x,y,z,t) = i\hbar\frac{\partial}{\partial t}\psi(x,y,z,t)$$

And here's the 3D Schrödinger equation using the Laplacian, which feels a bit tidier:

$$\frac{-\hbar^2}{2m}\nabla^2\psi(x,y,z,t) + V(x,y,z,t)\psi(x,y,z,t) = i\hbar\frac{\partial}{\partial t}\psi(x,y,z,t)$$

To solve this equation, when the potential doesn't vary with time, break out the time-dependent part of the wave function:

$$\psi(x,y,z,t) = \psi(x,y,z)e^{-iEt/\hbar}$$

Here, $\psi(x, y, z)$ is the solution of the time-independent Schrödinger equation, and E is the energy:

$$\frac{-\hbar^2}{2m}\nabla^2\psi(x,y,z) + V(x,y,z)\psi(x,y,z) = E\psi(x,y,z)$$

So far, so good. But now you've run into a wall. The expression for the Laplacian, $\nabla^2\psi(x, y, z)$, is hard to deal with, so the current equation is — in general — hard to solve.

Separating the Schrödinger equation

So, what should you do with the difficult-to-solve equation from the previous section? You can focus on the case in which the equation is *separable* — that is, where you can separate out the x, y, and z dependence and find the solution in each dimension separately. In other words, in separable cases, the potential, $V(x, y, z)$, is actually the sum of the x, y, and z potentials:

$$V(x, y, z) = V_x(x) + V_y(y) + V_z(z)$$

Now you can break the Hamiltonian in the time-dependent Schrödinger equation into three Hamiltonians, H_x, H_y, and H_z:

$$(H_x + H_y + H_z)\psi(x, y, z) = E\psi(x, y, z)$$

where

» $H_x = \dfrac{-\hbar^2}{2m}\dfrac{\partial^2}{\partial x^2} + V_x(x)$

» $H_y = \dfrac{-\hbar^2}{2m}\dfrac{\partial^2}{\partial y^2} + V_y(y)$

» $H_z = \dfrac{-\hbar^2}{2m}\dfrac{\partial^2}{\partial z^2} + V_z(z)$

When you divide the Hamiltonian, as in $(H_x + H_y + H_z)\psi(x, y, z) = E\psi(x, y, z)$, you can also divide the wave function that solves that equation. In particular, you can break the wave function into three parts, one each for x, y, and z:

$$\psi(x, y, z) = X(x)Y(y)Z(z)$$

where $X(x)$, $Y(y)$, and $Z(z)$ are functions of the coordinates x, y, and z and are not to be confused with the position operators. This separation of the wave function into three parts makes your calculations considerably easier, because now you can break the Hamiltonian into three separate operators added together, and you also get three separate energy components:

$$\left(\frac{-\hbar^2}{2m}\frac{\partial^2}{\partial x^2}+V_x(x)\right)+\left(\frac{-\hbar^2}{2m}\frac{\partial^2}{\partial y^2}+V_y(y)\right)+\left(\frac{-\hbar^2}{2m}\frac{\partial^2}{\partial z^2}+V_z(z)\right)=H$$

$$E=E_x+E_y+E_z$$

You now have three independent (and also time-independent) Schrödinger equations for the three dimensions:

» $$\frac{-\hbar^2}{2m}\frac{\partial^2}{\partial x^2}X(x)+V(x)X(x)=E_xX(x)$$

» $$\frac{-\hbar^2}{2m}\frac{\partial^2}{\partial y^2}Y(y)+V(y)Y(y)=E_yY(y)$$

» $$\frac{-\hbar^2}{2m}\frac{\partial^2}{\partial z^2}Z(z)+V(z)Z(z)=E_zZ(z)$$

TIP

This system of independent differential equations looks a lot easier to solve than $(H_x+H_y+H_z)\psi(x,y,z)=E\psi(x,y,z)$. In essence, you've broken the three-dimensional Schrödinger equation into three one-dimensional Schrödinger equations. That makes solving 3D problems tractable.

Solving 3D Free Particle Problems

Consider the free particle that's shown in three dimensions in Figure 12-1.

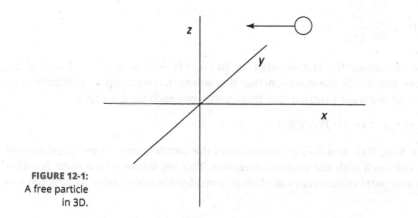

FIGURE 12-1:
A free particle
in 3D.

To solve the problem for the free particle shown in the figure, you can apply these steps (introduced in Chapter 8):

1. **Determine your coordinates to write the Schrödinger equation.**

   The heavy lifting for this part comes in the previous section. This chapter is, after all, about rectangular coordinates, so it's not surprising that I'm using rectangular coordinates to write the Schrödinger equation here. Then, separating the coordinates creates three one-dimensional Schrödinger equations.

2. **Apply specific constraints; simplify if possible.**

   Because the particle is traveling freely, $V(x) = V(y) = V(z) = 0$. There is no energy potential affecting the motion of the particle. The three independent Schrödinger equations for the three dimensions (covered in the preceding section) become the following:

   - $$\frac{-\hbar^2}{2m}\frac{\partial^2}{\partial x^2}X(x) = E_x X(x)$$

   - $$\frac{-\hbar^2}{2m}\frac{\partial^2}{\partial y^2}Y(y) = E_y Y(y)$$

   - $$\frac{-\hbar^2}{2m}\frac{\partial^2}{\partial z^2}Z(z) = E_z Z(z)$$

3. **Restructure the Schrödinger equation into a solvable form.**

   If you rewrite these equations in terms of the wave number, k, where $k^2 = \frac{2mE}{\hbar^2}$, then these equations become the following:

   - $$\frac{\partial^2}{\partial x^2}X(x) = -k_x{}^2 X(x)$$

   - $$\frac{\partial^2}{\partial y^2}Y(y) = -k_y{}^2 Y(y)$$

   - $$\frac{\partial^2}{\partial z^2}Z(z) = -k_z{}^2 Z(z)$$

4. **Solve the differential equations for ψ.**

   In this case, you have the wave function, ψ, in its separated form, as independent components X, Y, and Z. You solve for each of these components and then combine those solutions to get ψ.

REMEMBER

The coordinate functions $X(x)$, $Y(y)$, and $Z(z)$ are represented by a trio of second-order differential equations. Again, keep in mind that these are coordinate functions, and are *not* the position operators.

You can write the general solutions as

$$\frac{\partial^2}{\partial x^2}X(x) = -k_x^{\;2}X(x) \quad \rightarrow \quad X(x) = A_x e^{ik_x x}$$

$$\frac{\partial^2}{\partial y^2}Y(y) = -k_y^{\;2}Y(y) \quad \rightarrow \quad Y(y) = A_y e^{ik_y y}$$

$$\frac{\partial^2}{\partial z^2}Z(z) = -k_z^{\;2}Z(z) \quad \rightarrow \quad Z(z) = A_z e^{ik_z z}$$

where A_x, A_y, and A_z are constants.

Because $\psi(x, y, z) = X(x)Y(y)Z(z)$, you get this equation:

$$\psi(x, y, z) = A_x A_y A_z e^{ik_x x} e^{ik_y y} e^{ik_z z}$$

$$= A e^{i(k_x x + k_y y + k_z z)}$$

where $A = A_x A_y A_z$.

The part of the exponent in the parentheses is the *dot product* of the vectors **k** and **r**, **k·r**. If the vector $\boldsymbol{a} = (a_x, a_y, a_z)$ in terms of components and the vector $\boldsymbol{b} = (b_x, b_y, b_z)$, then the dot product of $\boldsymbol{a}$ and $\boldsymbol{b}$ is $\boldsymbol{a} \cdot \boldsymbol{b} = a_x b_x + a_y b_y + a_z b_z$. The dot product of two vectors is a scalar value.

Here's how you can rewrite the $\psi(x, y, z)$ wave function equation:

$$\psi(x, y, z) = A e^{i\boldsymbol{k} \cdot \boldsymbol{x}}$$

5. Use the boundary condition and normalization to find the constants.

Now that you have the wave function, you'll want to find the constants. You find out about this step in the section "Finding a physical solution," later in the chapter. But first, I give you a look at the energy in this situation.

Finding the total energy equation

The total energy of the free particle is the sum of the energy in three dimensions:

$$E = E_x + E_y + E_z$$

With a free particle, the energy of the x component of the wave function is $\frac{\hbar^2 k_x^{\;2}}{2m} = E_x$. And this equation works the same way for the y and z components, so here's the total energy of the particle:

$$E = \frac{\hbar^2 k_x^2}{2m} + \frac{\hbar^2 k_y^2}{2m} + \frac{\hbar^2 k_z^2}{2m}$$

$$= \frac{\hbar^2}{2m}\left(k_x^2 + k_y^2 + k_z^2\right)$$

Note that $k_x^2 + k_y^2 + k_z^2$ is the square of the magnitude of k — that is, $k \cdot k = k^2$. Therefore, you can write the equation for the total energy as

$$E = \frac{\hbar^2}{2m}\left(k_x^2 + k_y^2 + k_z^2\right) = \frac{\hbar^2}{2m}k^2$$

REMEMBER

Because E is a constant no matter where the particle is pointed (which axis it's on), you will have multiple states of k that share the same energy state. In fact, an infinite number of energy states exist. All the eigenfunctions represented by the differential equations you find in Step 3 in the preceding section are infinitely degenerate as you vary k_x, k_y, and k_z.

Adding time dependence

You can add time dependence to the solution for $\psi(x, y, z)$, giving you $\psi(x, y, z, t) = \psi(x, y, z)e^{-iEt/\hbar}$ (similar to the time-dependent Schrödinger equation from Chapter 8). That equation gives you this form for $\psi(x, y, z, t)$:

$$\psi(x, y, z, t) = Ae^{i\left(k \cdot r - Et/\hbar\right)}$$

Because $\omega = \frac{E}{\hbar}$, the equation becomes

$$\psi(x, y, z, t) = Ae^{i(k \cdot r - \omega t)}$$

In fact, now that the right side of the equation is in terms of the radius vector r, you can make the left side match those terms:

$$\psi(r, t) = Ae^{i(k \cdot r - \omega t)}$$

Finding a physical solution

Now you get to the point where you have the wave function, and you can try to normalize it (by setting the sum of the probability function across all space equal to one) to find the constants. Unfortunately for setting the integral to equal to one, integrating this wave function gives you the following, where A is a constant:

$$\int_{-\infty}^{+\infty} \psi(r, t)\psi^*(r, t)d^3r = |A|^2 \int_{-\infty}^{+\infty} d^3r \to \infty$$

This integral diverges and you can't normalize $\psi(r, t)$ as I've written it. This wave function seems to give a solution that isn't physically possible! What do you do here to get a physical particle?

TIP

It may be useful at this point to peek at the section in Chapter 8 where I discuss the one-dimensional Schrödinger equation for a free particle, and then follow these steps:

1. Convert the wave function to a wave packet.

TIP

The key to solving this problem is to realize that if you have a number of solutions to the Schrödinger equation, then any linear combination of those solutions is also a solution. In other words, you add various wave functions together so that you get a *wave packet,* which is a collection of wave functions of the form $e^{ik \cdot r}$ such that

- The wave functions interfere constructively at one location.

- They interfere destructively (go to zero) at all other locations.

Look at the time-independent version:

$$\psi(r) = \sum_{n=1}^{\infty} \phi_n e^{ik \cdot r}$$

However, for a free particle, the energy states are not separated into distinct bands; the possible energies are continuous, so people write this summation as an integral:

$$\psi(r) = \frac{1}{(2\pi)^{3/2}} \int_{-\infty}^{+\infty} \phi(k) e^{ik \cdot r} d^3 k$$

2. Choose a form for ϕ.

What is $\phi(k)$? It's the three-dimensional analogue of $\phi(k)$ that you find in Chapter 8; that is, it's the amplitude of each component wave function. You can find $\phi(k)$ from the Fourier transform of $\psi_1(x) = Ae^{ik_1 x} + Be^{-ik_1 x}$ (where $x < 0$) like this:

$$\phi(k) = \frac{1}{(2\pi)^{3/2}} \int_{-\infty}^{+\infty} \psi(r) e^{-ik \cdot r} d^3 k$$

In practice, you choose $\phi(k)$ yourself. Look at an example, using the following form for $\phi(k)$, which is for a Gaussian wave packet (**Note:** The exponential part is what makes this a Gaussian wave form):

$$\phi(k) = \phi(k) = A\exp\left(\frac{-a^2 k^2}{4}\right)$$

where a and A are constants.

3. Normalize ϕ to determine constants.

You can begin by normalizing $\phi(\boldsymbol{k})$ to determine what A is. Here's how that works:

$$1 = \int_{-\infty}^{+\infty} |\phi(k)|^2 \, d^3k = |A|^2 \int_{-\infty}^{+\infty} \exp\left(\frac{-a^2}{2} k^2\right) d^3k$$

Performing the integration gives you

$$1 = |A|^2 \left(\frac{2\pi}{a^2}\right)^{3/2}$$

$$A = \left(\frac{a^2}{2\pi}\right)^{3/4}$$

which means that the wave function is

$$\psi(\boldsymbol{r}, t) = \frac{1}{(2\pi)^{3/2}} \left(\frac{a^2}{2\pi}\right)^{3/4} \int_{-\infty}^{+\infty} \exp\left(\frac{-a^2}{2} k^2\right) e^{i\boldsymbol{k}\cdot\boldsymbol{r}} \, d^3k$$

4. Integrate to find ψ.

You can evaluate this equation to give you the following, which is what the time-independent wave function for a Gaussian wave packet looks like in 3D:

$$\psi(\boldsymbol{r}, t) = \left(\frac{2}{\pi a^2}\right)^{3/4} \exp\left(\frac{-r^2}{a^2}\right)$$

Okay, that's how the solution looks when $V(r) = 0$. But can't you solve some problems when $V(r)$ is not equal to zero? Yep, you sure can. Check out the next section.

Getting Squared Away with 3D Rectangular Potentials

This section takes a look at a 3D potential that forms a box, as shown in Figure 12-2. You want to get the wave functions and the energy levels for the 3D potential wells, in a way that's analogous to the one-dimensional potential wells that are the focus of Chapter 8. And you can follow the steps from that chapter.

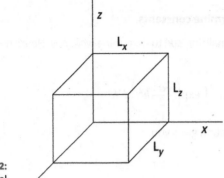

FIGURE 12-2:
A box potential
in 3D.

1. **Determine your coordinates to write the Schrödinger equation.**

2. **Apply specific constraints; simplify if possible.**

   Inside the box, say that V(x, y, z) = 0, and outside the box, say that V(x, y, z) = ∞. You have the following:

   $$V(x, y, z) = \begin{vmatrix} 0, \text{ where } 0 < x < L_x, \ 0 < y < L_y, \ 0 < z < L_z \\ \infty \text{ otherwise} \end{vmatrix}$$

   Dividing V(x, y, z) into $V_x(x)$, $V_y(y)$, and $V_z(z)$ gives you

   - $V_x(x) = \begin{vmatrix} 0, \text{ where } 0 < x < L_x \\ \infty \text{ otherwise} \end{vmatrix}$

   - $V_y(y) = \begin{vmatrix} 0, \text{ where } 0 < y < L_y \\ \infty \text{ otherwise} \end{vmatrix}$

   - $V_z(z) = \begin{vmatrix} 0, \text{ where } 0 < z < L_z \\ \infty \text{ otherwise} \end{vmatrix}$

   Because the potential goes to infinity at the walls of the box, the wave function, ψ(x, y, z), must go to zero at the walls, so that's your constraint. In 3D, the Schrödinger equation looks like this in three dimensions, and then you can write out the different components of the Laplacian ∇² to get an equivalent form:

   $$\frac{-\hbar^2}{2m} \nabla^2 \psi(x, y, z) + V(x, y, z)\psi(x, y, z) = E\psi(x, y, z)$$

3. **Restructure the Schrödinger equation into a solvable form.**

   Take this dimension by dimension. Because the potential is separable, you can write ψ(x, y, z) as ψ(x, y, z) = X(x)Y(y)Z(z). Inside the box, the potential equals zero. So the Schrödinger equation for x, y, and z will be identical within the box as if you were working with free particles, and you get the same three Schrödinger equations:

- $$\frac{\partial^2}{\partial x^2} X(x) = -k_x{}^2 X(x)$$

- $$\frac{\partial^2}{\partial y^2} Y(y) = -k_y{}^2 Y(y)$$

- $$\frac{\partial^2}{\partial z^2} Z(z) = -k_z{}^2 Z(z)$$

Remember that the wave number k is defined such that $k^2 = \dfrac{2mE}{\hbar^2}$.

4. **Solve the differential equation for ψ.**

Again, you can break the wave function ψ into separable parts, so you focus on one part and then extend that thinking to the other parts. Start by taking a look at the equation for x. Now you have something to work with — a second-order differential equation. Here are the two independent solutions to this equation, $X_1(x)$ and $X_2(x)$, where A and B are yet to be determined:

$$\frac{\partial^2}{\partial x^2} X(x) = -k_x{}^2 X(x) \rightarrow \begin{cases} X_1(x) = A \sin(kx) \\ X_2(x) = B \cos(kx) \end{cases}$$

The general solution is the sum of the last two equations:

$$X(x) = X_1(x) + X_2(x) = A \sin(kx) + B \cos(kx)$$

5. **Use the boundary condition and normalization to find constants.**

You have to use the boundary conditions to find the values of A and B. What are the boundary conditions? The wave function must disappear at the boundaries of the box, so

- $X(0) = 0$

- $X(L_x) = 0$

Trigonometry comes into play in solving for the constants, because of the sine and cosine. The fact that $\psi(0) = 0$ tells you right away that B must be 0, because $\cos(0) = 1$. I cover the rest of this calculation in the section "Normalizing the wave function," later in the chapter. But first, I take another brief look at the energy levels.

Determining the energy levels

From the boundary conditions in the previous section, the fact that $X(L_x) = 0$ tells you that $X(L_x) = A \sin(k_x L_x) = 0$. Because the sine is 0 when its argument is a multiple of π, this means that

$$k_x L_x = n_x \pi \qquad n_x = 1, 2, 3 \ldots$$

$$k_x = \frac{n_x \pi}{L_x}$$

And because $k^2 = \dfrac{2mE}{\hbar^2}$, it means that

$$\frac{2mE_x}{\hbar^2} = \frac{n_x^2 \pi^2}{L_x^2} \qquad n_x = 1, 2, 3 \ldots$$

$$E_x = \frac{n_x^2 \hbar^2 \pi^2}{2m\, L_x^2}$$

That's the energy in the x component of the wave function, corresponding to the quantum numbers 1, 2, 3, and so on. The total energy of a particle of mass m inside the box potential is $E = E_x + E_y + E_z$. Following $E_x = \dfrac{n_x^2 \hbar^2 \pi^2}{2m\, L_x^2}$, you have this for E_y and E_z:

$$E_y = \frac{n_y^2 \hbar^2 \pi^2}{2m\, L_y^2} \qquad n_y = 1, 2, 3 \ldots$$

$$E_z = \frac{n_z^2 \hbar^2 \pi^2}{2m\, L_z^2} \qquad n_z = 1, 2, 3 \ldots$$

The total energy of the particle is $E = E_x + E_y + E_z$, which equals this:

$$E = \frac{n_x^2 \hbar^2 \pi^2}{2m\, L_x^2} + \frac{n_y^2 \hbar^2 \pi^2}{2m\, L_y^2} + \frac{n_z^2 \hbar^2 \pi^2}{2m\, L_z^2}$$

for $n_x = 1, 2, 3 \ldots$, $n_y = 1, 2, 3 \ldots$, and $n_z = 1, 2, 3 \ldots$.

And there you have the total energy of a particle in the box potential.

Normalizing the wave function

Now how about normalizing the wave function $\psi(x, y, z)$? You can refer to Step 1 in the section "Getting Squared Away with 3D Rectangular Potentials." Take the equation from Step 4 and apply the constant $B = 0$ found in Step 5. You now have this for the wave equation in the x dimension:

$$X(x) = A \sin\left(\frac{n_x \pi x}{L_x}\right)$$

The wave function is a sine wave, going to zero at $x = 0$ and $x = L_z$. Technically, the following steps are a continuation of Step 5 in the referenced step list, but getting from there to the final wave function has a few stages. So, I walk you through the stages with a new set of steps, just to keep things clear.

1. **Normalize X to solve for A.**

   You can also insist that the wave function be normalized, like this:

   $$1 = \int_0^{L_x} |X(x)|^2 \, dx$$

   By normalizing the wave function, you can solve for the unknown constant A. Substituting for $X(x)$ in the equation gives you the following:

   $$1 = |A|^2 \int_0^{L_x} \sin^2\left(\frac{n_x \pi x}{L_x}\right) dx$$

   $$\int_0^{L_x} \sin^2\left(\frac{n_x \pi x}{L_x}\right) dx = \frac{L_x}{2}$$

   Therefore, $1 = |A|^2 \int_0^{L_x} \sin^2\left(\frac{n_x \pi x}{L_x}\right) dx$ becomes $1 = |A|^2 \frac{L_x}{2}$, which means you can solve for A:

   $$A = \sqrt{\frac{2}{L_x}}$$

   Great, now you have the constant A, so you can get $X(x)$:

   $$X(x) = \sqrt{\frac{2}{L_x}} \sin\left(\frac{n_x \pi x}{L_x}\right) \qquad n_x = 1, 2, 3 \ldots$$

2. **Repeat this process to find Y and Z.**

   Now get $\psi(x, y, z)$. You can divide the wave function into three parts:

   $\psi(x, y, z) = X(x)Y(y)Z(z)$

   By analogy with $X(x)$, you can find $Y(y)$ and $Z(z)$:

   $$Y(y) = \sqrt{\frac{2}{L_y}} \sin\left(\frac{n_y \pi y}{L_y}\right) \qquad n_y = 1, 2, 3 \ldots$$

   $$Z(z) = \sqrt{\frac{2}{L_z}} \sin\left(\frac{n_z \pi z}{L_z}\right) \qquad n_z = 1, 2, 3 \ldots$$

3. **Combine X, Y, and Z to get ψ.**

So, ψ(x, y, z) equals the following:

$$\psi(x, y, z) = \sqrt{\frac{8}{L_x L_y L_z}} \sin\left(\frac{n_x \pi x}{L_x}\right) \sin\left(\frac{n_y \pi x}{L_y}\right) \sin\left(\frac{n_z \pi x}{L_z}\right)$$

for $n_x = 1, 2, 3 \ldots$, $n_y = 1, 2, 3 \ldots$, and $n_z = 1, 2, 3 \ldots$.

That's a pretty long wave function. (If you've been following through the chapters in order, you may now understand why I choose to present one-dimensional problems whenever possible!) In fact, when you're dealing with a box potential, the energy looks like this:

$$E = \frac{n_x^2 \hbar^2 \pi^2}{2m\, L_x^2} + \frac{n_y^2 \hbar^2 \pi^2}{2m\, L_y^2} + \frac{n_z^2 \hbar^2 \pi^2}{2m\, L_z^2}$$

Using a cubic potential

When working with a box potential, you can make things simpler by assuming that the box is actually a cube. In other words, $L = L_x = L_y = L_z$. When the box is a cube, the equation for the energy becomes

$$E = \frac{\hbar^2 \pi^2}{2mL^2}(n_x^2 + n_y^2 + n_x^2)$$

for $n_x = 1, 2, 3 \ldots$, $n_y = 1, 2, 3 \ldots$, and $n_z = 1, 2, 3 \ldots$.

So, for example, the energy of the ground state, where $n_x = n_y = n_z = 1$, is given by the following, where E_{111} is the ground state:

$$E_{111} = \frac{3\hbar^2 \pi^2}{2mL^2}$$

Degenerate energies and symmetry

Note that some degeneracy exists in the energies. That is, you find various energy levels that will be identical. For example, note that

» E_{211} ($n_x = 2, n_y = 1, n_z = 1$) is $E_{211} = \dfrac{6\hbar^2 \pi^2}{2mL^2}$

» E_{121} ($n_x = 1, n_y = 2, n_z = 1$) is $E_{121} = \dfrac{6\hbar^2 \pi^2}{2mL^3}$

» E_{112} ($n_x = 1, n_y = 1, n_z = 2$) is $E_{112} = \dfrac{6\hbar^2 \pi^2}{2mL^2}$

So, $E_{211} = E_{121} = E_{112}$, which means that the first excited state is threefold degenerate, matching the threefold equivalence in dimensions.

REMEMBER

In general, when you have symmetry built into the physical layout (as you do when $L = L_x = L_y = L_z$), you have degeneracy. You have individual physical configurations that are identical to other physical configurations, so they also have identical energy values.

Cubic potential wave function

The wave function for a cubic potential is also easier to manage than the wave function for a general box potential (where the sides aren't of the same length). Here's the wave function for a cubic potential:

$$\psi(x, y, z) = \sqrt{\frac{8}{L^3}} \sin\left(\frac{n_x \pi x}{L}\right) \sin\left(\frac{n_y \pi y}{L}\right) \sin\left(\frac{n_z \pi z}{L}\right)$$

for $n_x = 1, 2, 3 \ldots$, $n_y = 1, 2, 3 \ldots$, and $n_z = 1, 2, 3 \ldots$.

So, for example, here's the wave function for the ground state ($n_x = 1$, $n_y = 1$, $n_z = 1$), $\psi_{111}(x, y, z)$:

$$\psi_{111}(x, y, z) = \sqrt{\frac{8}{L^3}} \sin\left(\frac{\pi x}{L}\right) \sin\left(\frac{\pi y}{L}\right) \sin\left(\frac{\pi z}{L}\right)$$

And here's $\psi_{211}(x, y, z)$:

$$\psi_{211}(x, y, z) = \sqrt{\frac{8}{L^3}} \sin\left(\frac{2\pi x}{L}\right) \sin\left(\frac{\pi y}{L}\right) \sin\left(\frac{\pi z}{L}\right)$$

And $\psi_{121}(x, y, z)$:

$$\psi_{121}(x, y, z) = \sqrt{\frac{8}{L^3}} \sin\left(\frac{\pi x}{L}\right) \sin\left(\frac{2\pi y}{L}\right) \sin\left(\frac{\pi z}{L}\right)$$

Springing into 3D Harmonic Oscillators

In one dimension, the general particle harmonic oscillator (which I describe in Chapter 9) looks like Figure 12-3, where the particle is under the influence of a restoring force — illustrated here as a spring.

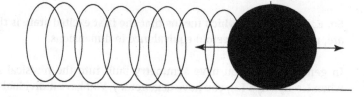

FIGURE 12-3:
A harmonic
oscillator.

Potential of a 3D spring

The restoring force has the form $F_x = -k_x x$ in one dimension, where k_x is the constant of proportionality between the force on the particle and the location of the particle. The potential energy of the particle as a function of location x is $V(x) = \frac{1}{2}k_x x^2$. This is also sometimes written as

$$V(x) = \frac{1}{2}m\omega_x^2 x^2$$

where $\omega_x^2 = \frac{k_x}{m}$.

In this section, you look at the harmonic oscillator in three dimensions. In three dimensions, the potential looks like this:

$$V(x, y, z) = \frac{1}{2}m\omega_x^2 x^2 + \frac{1}{2}m\omega_y^2 y^2 + \frac{1}{2}m\omega_z^2 z^2$$

$$\omega_x^2 = \frac{k_x}{m}, \qquad \omega_y^2 = \frac{k_y}{m}, \qquad \omega_z^2 = \frac{k_z}{m}$$

Solving Schrödinger for the 3D spring

Now that you have a form for the potential, you can start talking in terms of Schrödinger's equation:

$$\frac{-\hbar^2}{2m}\left(\frac{\partial^2}{\partial x^2} + \frac{\partial^2}{\partial y^2} + \frac{\partial^2}{\partial z^2}\right)\psi(x, y, z) + V(x, y, z)\psi(x, y, z) = E\psi(x, y, z)$$

Substituting in for the three-dimension potential, $V(x, y, z)$, gives you this equation:

$$\frac{-\hbar^2}{2m}\left(\frac{\partial^2}{\partial x^2} + \frac{\partial^2}{\partial y^2} + \frac{\partial^2}{\partial z^2}\right)\psi(x, y, z) + \left(\frac{1}{2}m\omega_x^2 x^2 + \frac{1}{2}m\omega_y^2 y^2 + \frac{1}{2}m\omega_z^2 z^2\right)$$
$$\psi(x, y, z) = E\psi(x, y, z)$$

Take this dimension by dimension. Because you can separate the potential into three dimensions, you can write $\psi(x, y, z)$ as $\psi(x, y, z) = X(x)Y(y)Z(z)$. Therefore, the Schrödinger equation looks like this for x:

$$\frac{-\hbar^2}{2m}\frac{\partial^2}{\partial x^2}X(x) + \frac{1}{2}m\omega_x^2 x^2 X(x) = E_x X(x)$$

You solve that equation for one dimension in Chapter 9, where you get this next solution:

$$X(x) = \frac{1}{\pi^{1/4}\sqrt{2^{n_x} n_x! x_0}} H_{n_x}\left(\frac{x}{x_0}\right)\exp\left(\frac{-x^2}{2x_0^2}\right)$$

where $x_0 = \sqrt{\dfrac{\hbar}{m\omega_x}}$ and $n_x = 0, 1, 2,$ and so on.

The H_{nx} term indicates a hermite polynomial, introduced in Chapter 9. You can write the wave function like this:

$$\psi(x, y, z) = \frac{1}{\pi^{3/4}}\frac{1}{\sqrt{2^{n_x+n_y+n_z} n_x! n_y! n_z! x_0 y_0 z_0}} H_{n_x}\left(\frac{x}{x_0}\right)H_{n_y}\left(\frac{y}{y_0}\right)$$

$$H_{n_z}\left(\frac{z}{z_0}\right)\exp\left(-\frac{x^2}{2x_0^2} - \frac{y^2}{2y_0^2} - \frac{z^2}{2z_0^2}\right)$$

That's a relatively easy form for a wave function (although it might not look like it), and it's all made possible by the fact that you can separate the potential into three dimensions.

Energy of the 3D oscillator

What about the energy of the harmonic oscillator? The energy of a one-dimensional harmonic oscillator is $E = \left(n + \dfrac{1}{2}\right)\hbar\omega$. And by analogy, the energy of a three-dimensional harmonic oscillator is given by

$$E = \left(n_x + \frac{1}{2}\right)\hbar\omega_x + \left(n_y + \frac{1}{2}\right)\hbar\omega_y + \left(n_z + \frac{1}{2}\right)\hbar\omega_z$$

Now consider a case where $\omega_x = \omega_y = \omega_z = \omega$. Here, the direction of the measurement doesn't matter, so it's called an *isotropic harmonic oscillator*. The energy looks like this in that situation:

$$E = \left(n_x + n_y + n_z + \frac{3}{2}\right)\hbar\omega$$

As for the cubic potential, the energy of a 3D isotropic harmonic oscillator is degenerate. For example, $E_{112} = E_{121} = E_{211}$. In fact, it's possible to have more than threefold degeneracy for a 3D isotropic harmonic oscillator — for example, $E_{200} = E_{020} = E_{002} = E_{110} = E_{101} = E_{011}$.

In general, the degeneracy of a 3D isotropic harmonic oscillator is

$$\text{Degeneracy} = \frac{1}{2}(n+1)(n+2)$$

where $n = n_x + n_y + n_z$.

IN THIS CHAPTER

» **Opting for spherical coordinates to solve problems**

» **Writing equations for central potentials**

» **Locating free particles in spherical coordinates**

» **Dealing with square well potentials**

» **Figuring out isotropic harmonic oscillators**

Chapter **13**

Solving Spherical Coordinate Problems

You're probably familiar with latitude and longitude — coordinates that basically name a couple of angles as measured from the center of the Earth. Put together the angle east or west, the angle north or south, and the all-important distance from the center of the Earth, and you have a vector that gives a good description of location in three dimensions. That vector is part of a spherical coordinate system.

Navigators talk more about the pair of angles than the distance ("Earth's surface" is generally specific enough), but quantum physicists find both angles and radius length important. Some 3D quantum physics problems even allow you to break down a wave function into an angular part and a radial part.

In this chapter, I discuss three-dimensional problems that are best handled using spherical coordinates. (For 3D problems that work better in rectangular coordinate systems, see Chapter 12.)

Note: When working through solutions in this chapter, you put into use your math skills related to the use of spherical coordinates, specifically as they apply in multivariable calculus and differential equations. And, since we're dealing with spherical coordinates, being familiar with trigonometry will help when the sines and cosines show up.

Choosing Spherical Coordinates

In Chapter 12, I discuss a 3D box potential and the potential well that traps the particle. You use rectangular coordinates in that case, because the 3D box potential is a rectangular prism and the potential is defined in each of the three coordinate directions. You can even simplify by considering the case where the well is a cube, and the potential is the same in each direction.

But what if the potential well that traps a particle has spherical symmetry, and is not rectangular or cubic? For example, suppose that the potential well looked like the following equation, where r is the radius of the particle's location with respect to the origin and a is a constant:

$$V(r) = \begin{cases} 0, \text{ where } 0 < r < a \\ \infty \text{ otherwise} \end{cases}$$

Trying to stuff this kind of problem into a rectangular-coordinates kind of solution is only asking for trouble, because — although you can solve it that way — doing so involves lots of sines and cosines, and results in a pretty complex solution. A much better tactic is to solve this kind of problem in the natural coordinate system in which the potential is expressed: spherical coordinates.

REMEMBER

Figure 13-1 shows the spherical coordinate system along with the corresponding rectangular coordinates, x, y, and z. In the spherical coordinate system, you locate points with a radius vector named r, which has three components.

>> r: The length of the radius vector.

>> θ: The angle from the z-axis to the r vector.

>> ϕ: The angle from the x-axis to the projection of the r vector in the x-y plane.

TIP

When you use spherical coordinates to solve problems that are *best expressed in spherical coordinates* (like a radially symmetric potential), you don't really have to convert between spherical coordinates and rectangular coordinates. If you do want to be prepared for doing that, you can look to the information in Chapter 10.

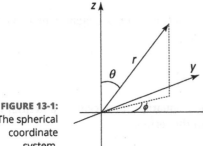

FIGURE 13-1:
The spherical
coordinate
system.

Observing Central Potentials in 3D

This chapter focuses on problems that involve central potentials. In these spherically symmetrical potentials, of the kind where $V(r) = V(r)$, the potential depends on only the magnitude of vector r (which is r), not on the angle of the vector.

When you work on problems that have a central potential, you can separate the wave function into a radial part (which depends on the form of the potential) and an angular part, which is a spherical harmonic. If you're not sure that you know how to handle harmonics with quantum physics, then you may want to check out Chapter 9. Quantum harmonic behavior is one of the types of problems that quantum physics is very good at dealing with!

Breaking down the Schrödinger equation

The Schrödinger equation looks like the following in three dimensions, where ∇^2 is the Laplacian operator (see Chapter 7 for more on operators and an introduction to the Schrödinger equation):

$$\frac{-\hbar^2}{2m}\nabla^2\psi(r)+V(r)\psi(r)=E\psi(r)$$

The Laplacian is a *differential operator* (it differentiates the thing it operates on), but you can apply it in a way that makes sense to the coordinates you are using. Throughout the book, I mostly work with the Laplacian operator in rectangular coordinates. In spherical coordinates, it's a little messy, but you can simplify it later. Check out the spherical Laplacian operator:

$$\nabla^2=\frac{1}{r}\frac{\partial^2}{\partial r^2}r-\frac{1}{\hbar^2 r^2}L^2$$

In this equation, L^2 is the square of the *orbital angular momentum* (which, sadly, is also more than a little messy):

$$L^2 = -\hbar^2\left(\frac{1}{\sin\theta}\frac{\partial}{\partial\theta}\left(\sin\theta\frac{\partial}{\partial\theta}\right) + \frac{1}{\sin^2\theta}\frac{\partial^2}{\partial\phi^2}\right)$$

In spherical coordinates, the Schrödinger equation for a central potential looks like the following when you substitute in the terms:

$$\frac{-\hbar^2}{2m}\frac{1}{r}\frac{\partial^2}{\partial r^2}r\psi(r) + \frac{1}{2mr^2}L^2\psi(r) + V(r)\psi(r) = E\psi(r)$$

The first term in the preceding equation actually corresponds to the radial kinetic energy — that is, the kinetic energy of the particle moving in the radial direction. The second term corresponds to the rotational kinetic energy. And the third term corresponds to the *potential energy*.

What can you say about the solutions to this version of the Schrödinger equation? Try looking at each term in this equation separately.

» $\frac{-\hbar^2}{2m}\frac{1}{r}\frac{\partial^2}{\partial r^2}r\psi(r)$: This first term is based entirely on the distance, r, and the second derivative is related to the change in that distance. This term looks at the change in the radial direction, movement either toward or away from the center of the potential well. Overall, this term corresponds to the *radial kinetic energy*. It is completely independent of the angles.

» $\frac{1}{2mr^2}L^2\psi(r)$: This second term is defined by the L^2, which I describe earlier in this section as the square of the orbital angular momentum. The orbital kinetic energy of the wave function is the *rotational kinetic energy*, which is inherently dependent entirely on the angular change.

» $V(r)\psi(r)$: The third term is the *potential energy*. In the case of this central potential, as I also note earlier, the potential energy is based only on the radius, r. Like the first term, this term is completely independent of the angles.

You can note that the first and third terms depend only on r and that the second term depends only on angles. This special property of problems with central potentials allows you to break their wave functions into a radial part and angular part.

The angular part of $\psi(r, \theta, \phi)$

When you have a central potential, what can you say about the angular part of $\psi(r, \theta, \phi)$? The angular part must be an eigenfunction of L^2, and as I show in Chapter 10, the eigenfunctions of L^2 are the spherical harmonics, $Y_{\ell m}(\theta, \phi)$ (where

ℓ is the total angular momentum quantum number and m is the z component of the angular momentum's quantum number). The angular part of the wave function is a spherical harmonic.

The radial part of $\psi(r, \theta, \phi)$

You can give the radial part of the wave function the name $R_{n\ell}(r)$, where n is a quantum number corresponding to the quantum state of the radial part of the wave function and ℓ is the total angular momentum quantum number. The radial part is symmetric with respect to angles, so it can't depend on m, the quantum number of the z component of the angular momentum. In other words, the wave function for particles in central potentials looks like the following equation in spherical coordinates:

$$\psi(r, \theta, \phi) = R_{n\ell}(r)Y_{\ell m}(\theta, \phi)$$

The next step is to solve for $R_{n\ell}(r)$ in general. Substituting $\psi(r, \theta, \phi)$ from the preceding equation into the Schrödinger equation, $\dfrac{-\hbar^2}{2m}\dfrac{1}{r}\dfrac{\partial^2}{\partial r^2}r\psi(r) + \dfrac{1}{2mr^2}L^2\psi(r) + V(r)\psi(r) = E\psi(r)$, gives you

$$-\hbar^2 \frac{r}{R_{n\ell}(r)}\frac{d^2}{dr^2}[r\,R_{n\ell}(r)] + 2mr^2[V(r) - E] + \frac{L^2Y_{\ell m}(\theta, \phi)}{Y_{\ell m}(\theta, \phi)} = 0$$

Note (from Chapter 10) that the spherical harmonics are eigenfunctions of L^2 (that's the whole reason for using them), with eigenvalue $\ell(\ell+1)\hbar^2$:

$$L^2Y_{\ell m}(\theta, \phi) = \ell(\ell+1)\hbar^2 Y_{\ell m}(\theta, \phi)$$

You can plug this eigenvalue into the last term of the preceding version of the Schrödinger equation to simplify it a bit, as follows:

$$\frac{L^2Y_{\ell m}(\theta, \phi)}{Y_{\ell m}(\theta, \phi)} = \frac{\ell(\ell+1)\hbar^2 Y_{\ell m}(\theta, \phi)}{Y_{\ell m}(\theta, \phi)} = \ell(\ell+1)\hbar^2$$

Substitute the simplified eigenvalue back into the Schrödinger equation to get

$$-\hbar^2 \frac{r}{R_{n\ell}(r)}\frac{d^2}{dr^2}[r\,R_{n\ell}(r)] + 2mr^2[V(r) - E] + \ell(\ell+1)\hbar^2 = 0$$

Then rewrite it as

$$\frac{-\hbar^2}{2m}\frac{d^2}{dr^2}[r\,R_{n\ell}(r)] + \left[V(r) + \frac{\ell(\ell+1)\hbar^2}{2mr^2}\right][r\,R_{n\ell}(r)] = E[r\,R_{n\ell}(r)]$$

The preceding equation is the one you use to determine the radial part of the wave function, $R_{n\ell}(r)$. It's called the radial equation for a central potential.

When you solve the radial equation for $R_{n\ell}(r)$, you can then find $\psi(r, \theta, \phi)$ because you already know $Y_{\ell m}(\theta, \phi)$:

$$\psi(r, \theta, \phi) = R_{n\ell}(r) Y_{\ell m}(\theta, \phi)$$

The radial equation is really a differential equation in one dimension: the r dimension. By selecting only problems that contain central potentials, you reduce the general problem of finding the wave function of particles trapped in a three-dimensional spherical potential to a one-dimensional differential equation.

Handling Free Particles in 3D with Spherical Coordinates

In this section and the next, you take a look at some example central potentials to see how to solve the radial equation (see the preceding section for more on the radial part). Here, you work with a free particle in a situation where no potential at all constrains the particle.

The wave function in spherical coordinates takes this form:

$$\psi(r, \theta, \phi) = R_{n\ell}(r) Y_{\ell m}(\theta, \phi)$$

And you know all about $Y_{\ell m}(\theta, \phi)$, because it gives you the spherical harmonics. The problem is now to solve for the radial part, $R_{n\ell}(r)$. Here's the radial equation:

$$\frac{-\hbar^2}{2m} \frac{d^2}{dr^2}[r R_{n\ell}(r)] + \left[V(r) + \frac{\ell(\ell+1)\hbar^2}{2mr^2} \right][r R_{n\ell}(r)] = E[r R_{n\ell}(r)]$$

Your goal is to solve the radial equation for the radial part of the wave function, $R_{n\ell}(r)$, so you can then combine it with the angular part, $Y_{\ell m}(\theta, \phi)$, to get the full wave function, $\psi(r, \theta, \phi)$. Just follow these steps to find the radial part of the wave function:

1. **Apply the potential constraint.**

   For a free particle, $V(r) = 0$, so the radial equation becomes

   $$\frac{-\hbar^2}{2m} \frac{d^2}{dr^2}[r R_{n\ell}(r)] + \frac{\ell(\ell+1)\hbar^2}{2mr^2}[r R_{n\ell}(r)] = E[r R_{n\ell}(r)]$$

Again, even in this simplified situation, you have another messy equation, and so you can take steps to clean it up a bit to make a more solvable problem.

2. Substitute to simplify.

Normally, in this free particle situation, you would define $k = (2mE)^{1/2}/\hbar$ to simplify the equation. In this radial equation, though, there's that variable r sitting around, so instead of substituting a constant, k, in this case you will substitute a variable, ρ.

You'll want to be clever in how you define ρ, though, so that it will create a form that will simplify the problem in Steps 4 and 5. The variable ρ is proportional to r, and the definition you will use is $\rho = kr$. This leads you to the following relationship:

$$\frac{\hbar^2}{2mr^2} = \frac{E}{\rho^2}$$

You can substitute this new relationship into the first two terms on the right of the radial equation, and now all of the terms will have the E variable. Now you can cancel E from every term in the equation, so you can solve it independently from the energy. That's one less variable to worry about!

3. Remove the n index.

Because a version of the same equation exists for each n index, you can conveniently remove the index so that $R_{n\ell}(r)$ becomes $R_{\ell}(\rho)$.

4. Set the radial equation equal to 0.

Substituting in the values from the previous steps, moving some things around with algebra, and performing the calculus results in the following radial equation:

$$\frac{d^2 R_{\ell}(\rho)}{d\rho^2} + \frac{2}{\rho}\frac{dR_{\ell}(\rho)}{d\rho} + \left[1 - \frac{\ell(\ell+1)}{\rho^2}\right]R_{\ell}(\rho) = 0$$

5. Solve for $R_{\ell}(\rho)$.

For the last step in this list, some special types of functions come to the rescue. You find out about those — the Bessel and Neumann functions — in the next section.

The spherical Bessel and Neumann functions

The radial part of the equation

$$\frac{d^2 R_\ell(\rho)}{d\rho^2} + \frac{2}{\rho}\frac{dR_\ell(\rho)}{d\rho} + \left[1 - \frac{\ell(\ell+1)}{\rho^2}\right]R_\ell(\rho) = 0$$

looks tough, but the solutions turn out to be well known. This equation is called the spherical Bessel equation, and the solution is a combination of the spherical Bessel functions [$j\ell(\rho)$] and the spherical Neumann functions [$n\ell(\rho)$]:

$$R_\ell(\rho) = A_\ell j_\ell(\rho) + B_\ell n_\ell(\rho)$$

where A_ℓ and B_ℓ are constants.

The spherical Bessel functions are given by the equation

$$j_\ell(\rho) = (-\rho)^\ell \left(\frac{1}{\rho}\frac{d}{d\rho}\right)^\ell \frac{\sin\rho}{\rho}$$

Here's what the first few iterations of $j_\ell(\rho)$ look like:

» $j_0(\rho) = \dfrac{\sin\rho}{\rho}$

» $j_1(\rho) = \dfrac{\sin\rho}{\rho^2} - \dfrac{\cos\rho}{\rho}$

» $j_2(\rho) = \dfrac{3\sin\rho}{\rho^3} - \dfrac{3\cos\rho}{\rho^2} - \dfrac{\sin\rho}{\rho}$

And the spherical Neumann functions are given by this equation:

$$n_\ell(\rho) = -(-\rho)^\ell \left(\frac{1}{\rho}\frac{d}{d\rho}\right)^\ell \frac{\cos\rho}{\rho}$$

Here are the first few iterations of $n_\ell(\rho)$:

» $n_0(\rho) = -\dfrac{\cos\rho}{\rho}$

» $n_1(\rho) = -\dfrac{\cos\rho}{\rho^2} - \dfrac{\sin\rho}{\rho}$

» $n_2(\rho) = -\dfrac{3\cos\rho}{\rho^3} - \dfrac{3\sin\rho}{\rho^2} + \dfrac{\cos\rho}{\rho}$

The limits for small and large ρ

According to the spherical Bessel equation (see the preceding section), the radial part of the wave function for a free particle looks like this:

$$R_\ell(\rho) = A_\ell j_\ell(\rho) + B_\ell n_\ell(\rho)$$

Take a look at the spherical Bessel functions and Neumann functions for small and large ρ.

>> Small ρ:

- The Bessel functions reduce to $j_\ell(\rho) \approx \dfrac{2^\ell \ell! \rho^\ell}{(2\ell + 1)!}$

- The Neumann functions reduce to $n_\ell(\rho) \approx \dfrac{-(2\ell - 1)! \rho^{-\ell-1}}{2^\ell \ell!}$

>> Large ρ:

- The Bessel functions reduce to $j_\ell(\rho) \approx \dfrac{1}{\rho} \sin\left(\rho - \dfrac{\ell\pi}{2}\right)$

- The Neumann functions reduce to $n_\ell(\rho) \approx -\dfrac{1}{\rho} \cos\left(\rho - \dfrac{\ell\pi}{2}\right)$

WARNING

The Neumann functions diverge for small values of ρ. Therefore, any wave function that includes the Neumann functions also diverges, which does not represent a real physical situation. So the Neumann functions aren't acceptable functions in the wave function. To avoid having the Neumann function influence, you can set $B_\ell = 0$, so that the Neumann functions aren't included in the wave function, leaving you with only

$$R_\ell(\rho) = A_\ell j_\ell(\rho)$$

Now that you have the radial component of the wave function simplified this far, you can combine that radial component with the angular components. You substitute $\rho = kr$ back in to express the wave function $\psi(r, \theta, \phi)$ as

$$\psi(r, \theta, \phi) = R_{n\ell}(\rho) Y_{\ell m}(\theta, \phi)$$
$$\psi(r, \theta, \phi) = A_\ell j_\ell(kr) Y_{\ell m}(\theta, \phi)$$

where $k = \dfrac{\sqrt{2mE_n}}{\hbar}$.

Note that because k can take any value, the energy levels are continuous.

Handling the Spherical Square Well Potential

Now consider the case of a spherical square well potential of the kind you can see in Figure 13-2. (I introduce square wells in Chapter 8.) This potential traps particles inside it. Mathematically, you can express the square well potential like this:

$$V(r) = \begin{cases} -V_0, \text{ where } 0 < r < a \\ 0, \text{ where } r > a \end{cases}$$

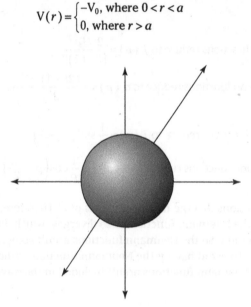

FIGURE 13-2:
The spherical
square well
potential.

Note that this potential is spherically symmetric and varies only in the radial r, and not in the angles θ or ϕ. You're dealing with a central potential, so you can break the wave function into an angular part and a radial part (see the section "Observing Central Potentials in 3D," earlier in the chapter).

TIP

Yes, I realize that I'm describing a situation that is spherical and calling it a "square well." As strange as it is, that's the terminology used in quantum physics, even when the square well is spherical in shape. Throughout the next few sections, I call this potential a "square well," but you and I both know that I'm not talking about something that's square, or even cubic, but instead, it's a sphere.

Looking inside the square well: 0 < r < a

For a spherical square well potential, here's what the radial equation looks like for the region 0 < r < a:

$$\frac{-\hbar^2}{2m}\frac{d^2}{dr^2}[r\,R_{n\ell}(r)]+\left[V(r)+\frac{\ell(\ell+1)\hbar^2}{2mr^2}\right][r\,R_{n\ell}(r)]=E[r\,R_{n\ell}(r)]$$

Here again, apply the same steps as in the preceding section for free particles.

1. **Apply the potential constraint.**

   In this region, V(r) = -V$_0$, so you have

   $$\frac{-\hbar^2}{2m}\frac{d^2}{dr^2}[r\,R_{n\ell}(r)]+\left[-V_0+\frac{\ell(\ell+1)\hbar^2}{2mr^2}\right][r\,R_{n\ell}(r)]=E[r\,R_{n\ell}(r)]$$

   Taking the V$_0$ term over to the right gives you the following:

   $$\frac{-\hbar^2}{2m}\frac{d^2}{dr^2}[r\,R_{n\ell}(r)]+\frac{\ell(\ell+1)\hbar^2}{2mr^2}[r\,R_{n\ell}(r)]=(E+V_0)r\,R_{n\ell}(r)$$

   In anticipation of the following steps, I can rewrite this equation slightly by multiplying every term by $-2m/r\hbar^2$ to simplify the equation:

   $$\frac{1}{r}\frac{d^2}{dr^2}[r\,R_{n\ell}(r)]-\frac{\ell(\ell+1)}{r^2}R_{n\ell}(r)=-\frac{2m}{\hbar^2}(E+V_0)R_{n\ell}(r)$$

   Doing so puts you in a better position to solve in the following steps.

2. **Substitute to simplify.**

   Now make the change of variable ρ = kr again, but with a different form of ρ than what was used for free particles. The particle is trapped in the square well, so its total energy is E + V$_0$, not just E. This results in the following relationship:

   $$\frac{\hbar^2}{2mr^2}=\frac{E+V_0}{\rho^2}$$

3. **Remove the n index.**

   This step is somewhat trivial, but you don't want to skip it. Because this equation exists for each n index, you can rewrite it without the index, and R$_{n\ell}$(r) becomes R$_\ell$(ρ).

4. Set the radial equation equal to 0.

Using this substitution, and getting all of the differentials of R grouped together, means that the radial equation takes the following form:

$$\frac{d^2R_\ell(\rho)}{d\rho^2} + \frac{2}{\rho}\frac{dR_\ell(\rho)}{d\rho} + \left[1 - \frac{\ell(\ell+1)}{\rho^2}\right]R_\ell(\rho) = 0$$

5. Solve for $R_\ell(\rho)$.

The solution to the preceding equation is a combination of the spherical Bessel functions $[j_\ell(\rho)]$ and the spherical Neumann functions $[n_\ell(\rho)]$ (just as you see for the free particle in the section "Handling Free Particles in 3D with Spherical Coordinates"):

$$R_\ell(\rho) = A_\ell j_\ell(\rho) + B_\ell n_\ell(\rho)$$

In this case, you also have a constraint on possible physical solutions: The wave function must be finite everywhere. This situation is confined to $0 < r < a$, and because $\rho = kr$, the only limits that you must check are for small values of ρ.

For small ρ, the Bessel functions look like this:

$$j_\ell(\rho) \approx \frac{2^\ell \ell! \rho^\ell}{(2^\ell+1)!}$$

And for small ρ, the Neumann functions reduce to

$$n_\ell(\rho) \approx \frac{-(2\ell-1)!\rho^{-\ell-1}}{2^\ell \ell!}$$

The Neumann functions diverge for small ρ, which makes them unacceptable for wave functions here. Ignoring the Neumann functions means that the radial part of the wave function is just made up of spherical Bessel functions, where A_ℓ is a constant:

$$R_\ell(\rho) = A_\ell j_\ell(\rho)$$

The whole wave function inside the square well, $\psi_{\text{inside}}(r, \theta, \phi)$, is a product of radial and angular parts, and it looks like this:

$$\psi_{\text{inside}}(r, \theta, \phi) = A_\ell j_\ell(\rho_{\text{inside}})Y_{\ell m}(\theta, \phi)$$

where $\rho_{\text{inside}} = \dfrac{r\sqrt{2m(E+V_0)}}{\hbar}$ and $Y_{\ell m}(\theta, \phi)$ are the spherical harmonics.

Outside the square well: $r > a$

Outside the square well, in the region $r > a$, the particle is just like a free particle. This is the region where $V(r) = 0$, after all. So here's what the radial equation looks like:

$$\frac{-\hbar^2}{2m}\frac{d^2}{dr^2}[r\,R_{n\ell}(r)] + \left[\frac{\ell(\ell+1)\hbar^2}{2mr^2}\right][r\,R_{n\ell}(r)] = E[r\,R_{n\ell}(r)]$$

You solve this equation in the section "Handling Free Particles in 3D with Spherical Coordinates," earlier in the chapter. Because $\rho = kr$, where $k = (2mE)^{1/2}/\hbar$, you substitute ρ for kr so that $R_{n\ell}(r)$ becomes $R_\ell(kr) = R_\ell(\rho)$. Using this substitution means that the radial equation takes the following form:

$$\frac{d^2R_\ell(\rho)}{d\rho^2} + \frac{2}{\rho}\frac{dR_\ell(\rho)}{d\rho} + \left[1 - \frac{\ell(\ell+1)}{\rho^2}\right]R_\ell(\rho) = 0$$

The solution is a combination of spherical Bessel functions and spherical Neumann functions, where B_ℓ and C_ℓ are constants:

$$R_\ell(r) = B_\ell j_\ell(\rho_{outside}) + C_\ell n_\ell(\rho_{outside})$$

TIP

When solving for free particles, the Neumann function was a nonphysical solution because it diverged for extremely small values of ρ. In this case, though, this equation applies only in the region $r > a$, so the Neumann function doesn't contribute any nonphysical infinities to the wave function. So, even though you removed the Neumann function for a free particle in empty space, you don't need to remove it for a free particle outside of a potential well.

So, the radial solution outside the square well looks like this, where $\rho_{outside} = r(2mE)^{1/2}/\hbar$:

$$\psi_{outside}(r, \theta, \phi) = [B_\ell j_\ell(\rho_{outside}) + C_\ell n_\ell(\rho_{outside})]Y_{\ell m}(\theta, \phi)$$

From the preceding section, you know that the wave function inside the square well is

$$\psi_{inside}(r, \theta, \phi) = A_l j_l(\rho_{inside})Y_{lm}(\theta, \phi)$$

So how do you find the constants A_ℓ, B_ℓ, and C_ℓ? You find those constants through continuity constraints: At the inside/outside boundary, where $r = a$, the wave function and its first derivative must be continuous. To determine A_ℓ, B_ℓ, and C_ℓ, you have to solve these two equations:

» $\psi_{inside}(a, \theta, \phi) = \psi_{outside}(a, \theta, \phi)$

» $\dfrac{d}{dr}\psi_{inside}(r, \theta, \phi)\Big|_{r=a} = \dfrac{d}{dr}\psi_{outside}(r, \theta, \phi)\Big|_{r=a}$

You now have the general solution for the wave function when you work with a spherical square well of potential energy. You can use these general solutions in specific situations, with specific quantum configurations, to find the missing constants. Have fun!

Getting the Goods on Isotropic Harmonic Oscillators

This section takes a look at *spherically symmetric* (that's the isotropic part) harmonic oscillators in three dimensions. Chapter 9 focuses on quantum harmonic oscillators in one dimension, so you may want to check out the relevant sections in that chapter before proceeding with the full 3D version.

In one dimension, you write the harmonic oscillator potential like this:

$$V(x) = \frac{1}{2} m\omega^2 x^2$$

where $\omega^2 = \frac{k}{m}$ (here, k is the spring constant; that is, the restoring force of the harmonic oscillator is $F = -kx$). You can turn these two equations into three-dimensional versions of the harmonic potential by replacing x with r:

$$V(r) = \frac{1}{2} m\omega^2 r^2$$

where $\omega^2 = \frac{k}{m}$. Because this potential is spherically symmetric, the wave function is going to be of the following form:

$$\psi(r, \theta, \phi) = R_{n\ell}(r) Y_{\ell m}(\theta, \phi)$$

where you have yet to solve for the radial function $R_{n\ell}(r)$ and where $Y_{\ell m}(\theta, \phi)$ describes the spherical harmonics.

As you may recognize, the radial Schrödinger equation looks like this:

$$\frac{-\hbar^2}{2m} \frac{d^2}{dr^2} [r\, R_{n\ell}(r)] + \left[V(r) + \frac{\ell(\ell+1)\hbar^2}{2mr^2} \right] [r\, R_{n\ell}(r)] = E[r\, R_{n\ell}(r)]$$

Substituting for V(r) from $V(r) = \frac{1}{2}m\omega^2 r^2$ gives you the following:

$$\frac{-\hbar^2}{2m}\frac{d^2}{dr^2}[r\,R_{n\ell}(r)] + \left[\frac{1}{2}m\omega^2 r^2 + \frac{\ell(\ell+1)\hbar^2}{2mr^2}\right][r\,R_{n\ell}(r)] = E[r\,R_{n\ell}(r)]$$

The solution to this equation is pretty difficult to obtain, and you're not going to gain anything by going through the math (pages and pages of it), so here's the solution:

$$R_{n\ell}(r) = C_{n\ell}r^\ell \exp\left(-m\omega\frac{r^2}{2\hbar}\right)L_n^{\ell+\frac{1}{2}}\left(m\omega\frac{r^2}{\hbar}\right)$$

where exp(x) = e^x and

$$C_{n\ell} = \frac{\left[\frac{2^{n+\ell+2}\left(\frac{m\omega}{\hbar}\right)^{\ell+\frac{3}{2}}}{\pi^{\frac{1}{2}}}\right]^{\frac{1}{2}}\left[\frac{n-\ell}{2}\right]!\left[\frac{n+\ell}{2}\right]!}{[(n+\ell+1)!]^{\frac{1}{2}}}$$

And the $L_a^b(r)$ functions are the generalized Laguerre polynomials, yet another type of polynomial you can look up that the mathematicians have helpfully developed to solve problems of this particular structure:

$$L_a^b(r) = \frac{r^{-b}e^r}{a!}\frac{d^a}{dr^a}\left(e^{-r}r^{a+b}\right)$$

Wow. Aren't you glad you didn't slog through the math? Here are the first few generalized Laguerre polynomials:

» $L_0^b(r) = 1$

» $L_1^b(r) = -r + b + 1$

» $L_2^b(r) = \frac{r^2}{2} - (b+2)r + \frac{(b+2)(b+1)}{2}$

» $L_3^b(r) = -\frac{r^3}{6} + \frac{(b+3)r^2}{2} - \frac{(b+2)(b+3)r}{2} + \frac{(b+1)(b+2)(b+3)}{6}$

All right, you have the form for $R_{n\ell}(r)$ from earlier in this section. To find the complete wave function, $\psi_{n\ell m}(r, \theta, \phi)$, you multiply by the spherical harmonics, $Y_{\ell m}(\theta, \phi)$:

$$\psi_{n\ell m}(r, \theta, \phi) = R_{n\ell}(r)Y_{\ell m}(\theta, \phi)$$

Now take a look at the first few wave functions for the isotropic harmonic oscillator in spherical coordinates:

$$\gg \quad \psi_{00m}(r,\theta,\phi) = \frac{2}{\pi^{\frac{1}{4}}}\left(\frac{m\omega}{\hbar}\right)^{\frac{3}{4}}\exp\left(-m\omega\frac{r^2}{2\hbar}\right)Y_{0m}(\theta,\phi)$$

$$\gg \quad \psi_{11m}(r,\theta,\phi) = \frac{\left(\frac{5}{3}\right)^{\frac{1}{2}}}{\pi^{\frac{1}{4}}}\left(\frac{m\omega}{\hbar}\right)^{\frac{5}{4}}r\left(5-\frac{2m\omega r^2}{\hbar}\right)\exp\left(-m\omega\frac{r^2}{2\hbar}\right)Y_{1m}(\theta,\phi)$$

$$\gg \quad \psi_{20m}(r,\theta,\phi) = \frac{1}{24^{\frac{1}{2}}\pi^{\frac{1}{4}}}\left(\frac{m\omega}{\hbar}\right)^{\frac{3}{4}}\left(15-20\frac{m\omega r^2}{\hbar}-4\frac{m^2\omega^2 r^4}{\hbar^2}\right)\exp\left(-m\omega\frac{r^2}{2\hbar}\right)Y_{0m}(\theta,\phi)$$

$$\gg \quad \psi_{21m}(r,\theta,\phi) = \pi^{\frac{3}{4}}\frac{3^{\frac{1}{2}}}{32}\left(\frac{m\omega}{\hbar}\right)^{\frac{5}{4}}r\left(35-28\frac{m\omega r^2}{\hbar}-4\frac{m^2\omega^2 r^4}{\hbar^2}\right)\exp\left(-m\omega\frac{r^2}{2\hbar}\right)Y_{1m}(\theta,\phi)$$

As you can see, when you have a potential that depends on r^2, as with harmonic oscillators, the wave function quickly gets extremely complex.

Fortunately, something relatively easy does come out of all of this. It turns out that the energy of an isotropic 3D harmonic oscillator is quantized, and you can derive the following relation for the energy levels:

$$E_n = \left(n+\frac{3}{2}\right)\hbar\omega \quad n=1,2,3\ldots$$

So, the energy levels start at $\frac{3\hbar\omega}{2}$ and then go to $\frac{5\hbar\omega}{2}, \frac{7\hbar\omega}{2}$, and so on.

This result is similar (though not identical) to the energy states found for one-dimensional harmonic oscillators in Chapter 9.

Chapter **14**

The Simplest Atom: Understanding Hydrogen

Not only is hydrogen the most common element in the universe, but it's also the simplest in terms of atomic structure, with just one proton and one electron. And one thing quantum physics is good at is predicting everything (such as energy levels and behavior) about simple atoms. This chapter is all about the hydrogen atom and solving the Schrödinger equation to find its energy. For such a little guy, the hydrogen atom can whip up a lot of math — and I help you to solve that math in this chapter.

Using the Schrödinger equation tells you just about all you need to know about the hydrogen atom, and its use is based on a single assumption: that the wave function must go to zero as r (the distance) goes to infinity. This assumption makes solving the Schrödinger equation possible. I start by introducing the Schrödinger equation for the hydrogen atom, and then I take you through calculating energy degeneracy and figuring out how far the electron is from the proton.

Note: In this chapter, you explore the wave function for a simple hydrogen atom. This exploration involves the Schrödinger equation, which you split up into the radial and angular components. I cover these two components extensively in Chapter 13, so make sure you're familiar with that content. In addition to the usual work related to solving the Schrödinger equation — calculus, linear algebra, and differential equations — this chapter also includes summations. In those cases, I skip over the calculations, but if you're a fan of summations, then you can definitely check them out on your own.

Revisiting Atomism

As I describe in Chapter 2, the idea of atoms significantly predates quantum physics. However, scientists in the nineteenth century tended to employ that idea as a useful tool to talk about different elements. They didn't necessarily believe that atoms actually existed. Even the creator of the periodic table of elements, Dmitri Mendeleev, was skeptical about whether the patterns of elements that showed up in his table meant that actual atoms exist.

The invention of quantum mechanics, however, brought the tools of science, and particularly of physics, directly to bear on questions about the atom. I cover the history of this research more extensively throughout Chapter 3 and Part 2 of this book, but in this chapter, I offer you Table 14-1, with milestones from those early years.

TABLE 14-1   **Early Atom-Related Milestones**

Date	Who	What Happened
1905	Albert Einstein	Explained Brownian motion (originally observed by Robert Brown in 1827) by using statistics to describe the motion of atoms in liquid form. This is largely seen as the solid experimental/theoretical evidence that led scientists to accept the existence of atoms.
1911	Ernest Rutherford	Created a model of the atom with a dense nucleus surrounded by electrons in orbit around it.
1913	Niels Bohr	Applied the tools of quantum physics to the Rutherford model of the atom. By quantizing energy states, he created the Bohr model of the hydrogen atom, which matched energy spectra experiments.

Date	Who	What Happened
1925	Werner Heisenberg	Created the matrix form of quantum mechanics (matrix mechanics).
	Wolfgang Pauli	Helped Heisenberg match matrix mechanics with the energy levels from the Bohr model of the hydrogen atom.
1926	Erwin Schrödinger	Created the wave function form of quantum mechanics (wave mechanics) and showed that it also predicted the energy levels from the Bohr model of the hydrogen atom.
	Jean-Baptiste Perrin	Received the Nobel Prize in Physics for his work confirming Einstein's explanation of Brownian motion, and for extensive other research into the "discontinuous structure of matter," showing that atoms and molecules exist.
1932	Werner Heisenberg	Received the Nobel Prize in Physics "for the creation of quantum mechanics, the application of which has, inter alia, led to the discovery of the allotropic forms of hydrogen."

From the timeline shown in the table, you can see that the investigation of atomic structure — and particularly of the atomic structure of hydrogen — was key in the early days and important creative moments of quantum physics. Physicists went on to use the tools of quantum physics to explore atomic structures more deeply, both in theoretical models and by providing the foundations for particle accelerators that allow them to probe these structures experimentally.

Coming to Terms: The Schrödinger Equation for the Hydrogen Atom

What do we now know about the hydrogen atom? In their basic description, hydrogen atoms are composed of a single proton, around which revolves a single electron. Figure 14-1 depicts this simple atomic structure.

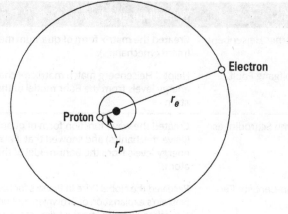

FIGURE 14-1:
The hydrogen
atom.

To walk you through getting the wave equation, I use the steps introduced in Chapter 8 (but in this chapter, you have to circle back and repeat some of the steps from different angles).

1. **Determine your coordinates to write the Schrödinger equation.**

REMEMBER

Refer to Figure 14-1 and notice that the proton isn't at the exact center of the atom — the center of mass shown as a solid black dot is at the exact center. In fact, the proton is at a radius of r_p from the exact center, and the electron is at a radius of r_e.

To get the wave equations you need, you construct the Schrödinger equation that fits this situation and includes terms for the kinetic and potential energy of the proton and the electron. Here are the terms.

● *For the proton's kinetic energy:*

$$\frac{-\hbar^2}{2m_p}\nabla_p^{\,2}$$

where $\nabla_p^{\,2} = \frac{\partial^2}{\partial x_p^{\,2}} + \frac{\partial^2}{\partial y_p^{\,2}} + \frac{\partial^2}{\partial z_p^{\,2}}$. In this expression, x_p is the proton's x position, y_p is the proton's y position, z_p is its z position, and m_p is the mass of the proton.

● *For the electron's kinetic energy:*

$$\frac{-\hbar^2}{2m_e}\nabla_e^{\,2}$$

where $\nabla_e^{\,2} = \frac{\partial^2}{\partial x_e^{\,2}} + \frac{\partial^2}{\partial y_e^{\,2}} + \frac{\partial^2}{\partial z_e^{\,2}}$. This term is identical to that of the kinetic energy of the proton, except that the variables all refer to the electron instead of the proton.

- *For the electron's kinetic energy:*

  The term for the potential energy in the Schrödinger equation will be just V(*r*), for now. Easy enough! I cover more specific constraints on that in Step 2.

  With these three terms defined, you then get the time-independent Schrödinger equation for this situation:

  $$\frac{-\hbar^2}{2m_p}\nabla_p^2\psi\left(r_e,\ r_p\right)-\frac{\hbar^2}{2m_e}\nabla_e^2\psi\left(r_e,\ r_p\right)+V\left(r\right)\psi\left(r_e,\ r_p\right)=E\psi\left(r_e,\ r_p\right)$$

  where $\psi(r_e,\ r_p)$ is the electron and proton's wave function.

2. **Apply specific constraints; simplify if possible.**

   The electrostatic potential energy, V(*r*), for a central potential is given by the following formula, where *r* is the radius vector separating the two charges:

   $$V\left(r\right)=-\frac{1}{4\pi\varepsilon_0}\frac{e^2}{|r|}$$

REMEMBER

   For problems like this, quantum physicists use the CGS (centimeter-gram-second) system of units, where $1=\frac{1}{4\pi\varepsilon_0}$. This is another one of those sneaky tricks, to simplify their calculations. In this system, the potential due to the electron and proton charges in the hydrogen atom is

   $$V\left(r\right)=\frac{-e^2}{|r|}$$

   Note that the distance between the electron and proton, *r*, can be written as $r=r_e-r_p$, so the preceding equation becomes

   $$V\left(r\right)=\frac{-e^2}{|r_e-r_p|}$$

   which gives you this Schrödinger equation:

   $$\frac{-\hbar^2}{2m_p}\nabla_p^2\psi\left(r_e,\ r_p\right)-\frac{\hbar^2}{2m_e}\nabla_e^2\psi\left(r_e,\ r_p\right)-\frac{e^2}{|r_e-r_p|}\psi\left(r_e,\ r_p\right)=E\psi\left(r_e,\ r_p\right)$$

Okay, so how do you handle this equation? Find out in the next section.

Simplifying and splitting the hydrogen equation

The quantum mechanical Schrödinger equation found in the previous section for the hydrogen atom is something of a mess, which helps explain why it was such a monster of a problem when physicists were first trying to understand it. The math is messy, because you must account for the distance of the proton from the atom's center of mass. Fortunately, the following steps can help you out.

1. **Restructure the Schrödinger equation into a solvable form.**

    You might think that you could simplify things by just assuming that the proton is stationary and that $r_p = 0$. This is a clever approach, but unfortunately, the proton does move around. As a result, using this particular trick gives you an equation that isn't exact.

    1a. *Switch to center-of-mass coordinates to simplify the usual Schrödinger equation.*

    The center of mass of the proton/electron system is at this location:

    $$\mathbf{R} = \frac{m_e \mathbf{r}_e + m_p \mathbf{r}_p}{m_e + m_p}$$

    And the vector between the electron and proton is $\mathbf{r} = \mathbf{r}_e - \mathbf{r}_p$.

    Using vectors $\mathbf{R}$ and $\mathbf{r}$ instead of $\mathbf{r}_e$ and $\mathbf{r}_p$ makes the Schrödinger equation easier to solve. The Laplacian for $\mathbf{R}$ is $\nabla_{\mathbf{R}}^2 = \frac{\partial^2}{\partial X^2} + \frac{\partial^2}{\partial Y^2} + \frac{\partial^2}{\partial Z^2}$. And the Laplacian for $\mathbf{r}$ is $\nabla_{\mathbf{r}}^2 = \frac{\partial^2}{\partial x^2} + \frac{\partial^2}{\partial y^2} + \frac{\partial^2}{\partial z^2}$.

    How can you relate $\nabla_{\mathbf{R}}^2$ and $\nabla_{\mathbf{r}}^2$ to the usual equation's ∇_p^2 and ∇_e^2? After the algebra settles, you get

    $$\frac{1}{m_e}\nabla_e^2 + \frac{1}{m_p}\nabla_p^2 = \frac{1}{M}\nabla_{\mathbf{R}}^2 + \frac{1}{m}\nabla_{\mathbf{r}}^2$$

    where $M = m_e + m_p$ is the total mass and $m = \frac{m_e m_p}{m_e + m_p}$ is called the *reduced mass*. When you put together the equations for the center of mass, the vector between the proton and the electron, the total mass, and m, then the time-independent Schrödinger equation becomes the following:

    $$\frac{-\hbar^2}{2M}\nabla_R^2 \psi(\mathbf{R}, \mathbf{r}) - \frac{\hbar^2}{2m}\nabla_r^2 \psi(\mathbf{R}, \mathbf{r}) + V(\mathbf{R}, \mathbf{r})\psi(\mathbf{R}, \mathbf{r}) = E\psi(\mathbf{R}, \mathbf{r})$$

    1b. *Adjust the potential terms in the Schrödinger equation.*

    Given the vectors, $\mathbf{R}$ and $\mathbf{r}$, the potential is given by

    $$V(\mathbf{R}, \mathbf{r}) = V(\mathbf{r}) = \frac{-e^2}{|\mathbf{r}|}$$

and the Schrödinger equation then becomes

$$\frac{-\hbar^2}{2M}\nabla_{\mathbf{R}}^2\psi(\mathbf{R},\mathbf{r}) - \frac{\hbar^2}{2m}\nabla_{\mathbf{r}}^2\psi(\mathbf{R},\mathbf{r}) - \frac{e^2}{|\mathbf{r}|}\psi(\mathbf{R},\mathbf{r}) = E\psi(\mathbf{R},\mathbf{r})$$

This equation looks easier — the main improvement being that you now have $|\mathbf{r}|$ in the denominator of the potential energy term rather than $|\mathbf{r}_e - \mathbf{r}_p|$.

1c. *Rewrite the equation as a separable differential equation.*

Because the equation contains terms involving either $\mathbf{R}$ or $\mathbf{r}$ but not both, the form of this equation indicates that it's a separable differential equation. You can look for a solution of the following form:

$\psi(\mathbf{R}, \mathbf{r}) = \psi(\mathbf{R})\psi(\mathbf{r})$

Substituting the preceding equation into the equation from Step 1b gives you the following:

$$\frac{-\hbar^2}{2M}\nabla_{\mathbf{R}}^2\psi(\mathbf{R})\psi(\mathbf{r}) - \frac{\hbar^2}{2m}\nabla_{\mathbf{r}}^2\psi(\mathbf{R})\psi(\mathbf{r}) - \frac{e^2}{|\mathbf{r}|}\psi(\mathbf{R})\psi(\mathbf{r}) = E\psi(\mathbf{R})\psi(\mathbf{r})$$

You might think that solving the equation looks harder than before. However, dividing this equation by $\psi(\mathbf{R})\psi(\mathbf{r})$ gives you

$$\frac{-\hbar^2}{2M\psi(\mathbf{R})}\nabla_{\mathbf{R}}^2\psi(\mathbf{R}) - \frac{\hbar^2}{2m\psi(\mathbf{r})}\nabla_{\mathbf{r}}^2\psi(\mathbf{r}) - \frac{e^2}{|\mathbf{r}|} = E$$

which is an equation with terms that depend on either $\psi(\mathbf{R})$ or $\psi(\mathbf{r})$ but not both. You can separate this equation into two equations, (where the total energy, E, equals $E_R + E_r$), and juggle them around a bit to get

- $\dfrac{-\hbar^2}{2M}\nabla_{\mathbf{R}}^2\psi(\mathbf{R}) = E_R\psi(\mathbf{R})$

- $\dfrac{-\hbar^2}{2m}\nabla_{\mathbf{r}}^2\psi(\mathbf{r}) - \dfrac{e^2}{|\mathbf{r}|}\psi(\mathbf{r}) = E_r\psi(\mathbf{r})$

2. Solve the differential equations for ψ.

You can solve the equations from Step 1c independently, which takes a bit of work.

2a. *Solve for ψ(R).*

In $\dfrac{-\hbar^2}{2M}\nabla_{\mathbf{R}}^2\psi(\mathbf{R}) = E_R\psi(\mathbf{R})$, how do you solve for $\psi(R)$, which is the wave function of the center of mass of the electron/proton system? This is a straightforward differential equation, and the solution is

$\psi(\mathbf{R}) = Ce^{-i\mathbf{k}\cdot\mathbf{r}}$

In this equation, C is a constant, and $\mathbf{k}$ is the wave vector where

$|\mathbf{k}| = \dfrac{2M\sqrt{E_R}}{\hbar^2}$.

REMEMBER

In practice, E_R is so small that people almost always just ignore $\psi(\mathbf{R})$ — that is, they assume it to be 1. In other words, the real action is in $\psi(\mathbf{r})$, not in $\psi(\mathbf{R})$; $\psi(\mathbf{R})$ is the wave function for the center of mass of the hydrogen atom, and $\psi(\mathbf{r})$ is the wave function for a (fictitious) particle of mass m.

2b. *Solve for $\psi(r)$.*

The Schrödinger equation for $\psi(\mathbf{r})$ is the wave function for a made-up particle of mass m. Here's the Schrödinger equation for $\psi(\mathbf{r})$:

$$\frac{-\hbar^2}{2m}\nabla_r^2\psi(\mathbf{r}) - \frac{e^2}{|\mathbf{r}|}\psi(\mathbf{r}) = E_r\psi(\mathbf{r})$$

TIP

In practice, $m \approx m_e$ and $\psi(\mathbf{r})$ is pretty close to $\psi(\mathbf{r}_e)$, so the energy, E_r, is pretty close to the electron's energy. In Step 1 of this section, I mention that you might think you can set $r_p = 0$ and treat the proton as if it were the stationary center of the hydrogen atom. If you try that with the original equation, you end up with an equation that looks very similar to this Schrödinger equation for $\psi(\mathbf{r})$. But that equation wouldn't be exact, for a couple of reasons. First, you'd be using the electron's mass, m_e, instead of the reduced mass, m. Second, it would be missing the slight contribution from $\psi(\mathbf{R})$.

Look for an exact equation for $\psi(\mathbf{r})$, with no cutting corners. You can break the solution, $\psi(\mathbf{r})$, into a radial part and an angular part (see Chapter 13 for more about this splitting):

$\psi(\mathbf{r}) = R_{n\ell}(r)Y_{\ell m}(\theta, \phi)$

The angular part of $\psi(\mathbf{r})$ is made up of spherical harmonics, $Y_{\ell m}(\theta, \phi)$, so that part's okay. Now you have to solve for the radial part, $R_{n\ell}(r)$. Here's what the Schrödinger equation becomes for the radial part:

$$\frac{-\hbar^2}{2m}\frac{d^2}{dr^2}[r\,R_{n\ell}(r)] + \ell(\ell+1)\frac{\hbar^2}{2mr^2}r\,R_{n\ell}(r) - \frac{e^2}{r}r\,R_{n\ell}(r) = E_r r\,R_{n\ell}(r)$$

where $r = |\mathbf{r}|$.

Solving the radial Schrödinger equation

To solve the final equation from the preceding section, you take a look at two cases — where r is very small and where r is very large. Putting them together gives you the rough form of the radial solution.

Looking at small r

For small r, the terms $-\frac{e^2}{r}r\,R_{n\ell}(r)$ and $E_r r R_{n\ell}(r)$, in the previous equation, become much smaller than the rest, so you can ignore them and write the radial Schrödinger as

$$\frac{-\hbar^2}{2m}\frac{d^2}{dr^2}[r\,R_{n\ell}(r)] + \ell(\ell+1)\frac{\hbar^2}{2mr^2}r\,R_{n\ell}(r) = 0$$

Then multiply by $2m\hbar^2$, and you get

$$\frac{-d^2}{dr^2}[r\,R_{n\ell}(r)] + \frac{\ell(\ell+1)}{r^2}r\,R_{n\ell}(r) = 0$$

The solution to this equation is proportional to $R_{n\ell}(r) \sim Ar^\ell + Br^{-\ell-1}$.

Note, however, that $R_{n\ell}(r)$ must vanish as r goes to zero — but the $r^{-\ell-1}$ term goes to infinity. And that means that B must be zero, so you have this solution for small r:

$$R_{n\ell}(r) \sim r^\ell$$

Looking at large r

For very large r, $\frac{-\hbar^2}{2m}\nabla_r^2\psi(r) - \frac{e^2}{|r|}\psi(r) = E_r\psi(r)$ becomes $\frac{d^2}{dr^2}[r\,R_{n\ell}(r)] + \frac{2mE_r}{\hbar^2}r\,R_{n\ell}(r) = 0$.

Because the electron is in a bound state in the hydrogen atom, E < 0; thus, the solution to the preceding equation is proportional to

$$R_{n\ell}(r) \sim Ae^{-\lambda r} + Be^{\lambda r}$$

where $\lambda = \frac{\sqrt{-2mE_r}}{\hbar}$.

Note: $R_{n\ell}(r) \sim Ae^{-\lambda r} + Be^{\lambda r}$ diverges as r goes to infinity because of the $Be^{\lambda r}$ term, so B must be equal to zero. That means that $R_{n\ell}(r) \sim e^{-\lambda r}$.

Looking at both r solutions

Putting together the solutions for small r and large r (from the previous sections), the Schrödinger equation gives you a solution to the radial Schrödinger equation of $R_{n\ell}(r) = r^\ell f(r)e^{-\lambda r}$, where $f(r)$ is some as-yet-undetermined function of r. Your next task is to determine f(r), which you do by substituting this equation into the radial Schrödinger equation, giving you the following:

$$\frac{-\hbar^2}{2m}\frac{d^2}{dr^2}[r\,R_{n\ell}(r)]+\ell(\ell+1)\frac{\hbar^2}{2mr^2}r\,R_{n\ell}(r)-\frac{e^2}{r}r\,R_{n\ell}(r)=E_r r\,R_{n\ell}(r)$$

Performing the substitution gives you the following differential equation:

$$\frac{d^2}{dr^2}f(r)+2\left[\frac{\ell+1}{r}-\lambda\right]\frac{df(r)}{dr}+2\left[\frac{\frac{me^2}{\hbar^2}-\lambda(\ell+1)}{r}\right]f(r)=0$$

Powering through the differential equation

The previous section leaves you with quite a differential equation, eh? You can solve it by using a power series (which is a common way of solving differential equations). By employing the correct power-series form of f(r) and skipping the two pages of calculations needed to simplify the series (you're welcome), you get the resulting function:

$$f(r)=\sum_{k=0}^{\infty}a_k r^k=e^{2\lambda r}$$

The radial wave function, $R_{n\ell}(r)$, looks like this:

$$R_{n\ell}(r)=r^\ell f(r)e^{-\lambda r}$$

where, as a reminder, $\lambda=\dfrac{\sqrt{-2mE}}{\hbar}$.

Plugging the form you have for $f(r)=e^{2\lambda r}$ into $R_{n\ell}(r)$ gives you the following:

» $R_{n\ell}(r)=r^\ell f(r)e^{-\lambda r}$

» $R_{n\ell}(r)=r^\ell e^{2\lambda r}e^{-\lambda r}$

» $R_{n\ell}(r)=r^\ell e^{\lambda r}$

Here's what the wave function $\psi(r)$ looks like: $\psi(r)=R_{n\ell}(r)\,Y_{\ell m}(\theta,\phi)$. And substituting in your form of $R_{n\ell}(r)$ from this equation gives you

$$\psi(r)=r^\ell e^{\lambda r}\,Y_{\ell m}(\theta,\phi)$$

WARNING

This final equation looks fine — *except* that as the exponent r increases, the value of the whole function will increase even faster! The wave function goes to infinity as r goes to infinity. But since it represents probability of a particle, you expect $\psi(r)$ to go to zero as r goes to infinity. (This assumption is explicitly stated at the start of the chapter.) This version of $R_{n\ell}(r) = r^\ell e^{\lambda r}$ is clearly unphysical. In other words, something went wrong somewhere. See the next section to find out how you fix this version of f(r) and, ultimately, $\psi(r)$.

Fixing *f(r)* to keep it finite

You need the solution for the radial equation (from the preceding section) to go to zero as r goes to infinity, which means fixing f(r) to keep it finite. This brings the problem to one of the peskiest steps.

1. **Set a finite boundary for the f(r) summation.**

   The problem of having $\psi(r)$ go to infinity as r goes to infinity lies in the form you assume for f(r) in the preceding section, which is

   $$f(r) = \sum_{k=0}^{\infty} a_k r^k$$

   The summation to infinity might give you pause, for good reason, because you have no reason to think that a hydrogen atom has infinitely many states. And you need to keep the number of states finite.

   To pull off this trick, the series must terminate at a certain index, which you call N. N is called the *radial quantum number*. So this equation becomes the following (note that the summation is now to N, and not to infinity):

   $$f(r) = \sum_{k=0}^{N} a_k r^k$$

2. **Establish constants equal to zero after the summation boundary.**

   For this series to terminate, every element after N must be equal to zero. This fact means that the constants for those terms ($a_{N+1}, a_{N+2}, a_{N+3}$, and so on) must all be zero. And this situation occurs when the following relationship holds:

   $$n\lambda - \frac{me^2}{\hbar^2} = 0 \quad n = 1, 2, 3 \dots$$

   In this expression, n is called the *principal quantum number*, and is calculated from $n = N + \ell + 1$. The series for f(r) must meet this quantization condition in order to be finite. And now only n quantum states are physically allowed.

   And because $\lambda = \dfrac{\sqrt{-2mE}}{\hbar}$, the equation

   $$n\lambda - \frac{me^2}{\hbar^2} = 0 \quad n = 1, 2, 3 \dots$$

   also puts constraints on the allowable values of the energy in the hydrogen atom.

Finding the allowed energies of the hydrogen atom

In the preceding section, you find that the quantization condition for $\psi(r)$ to remain finite as r goes to infinity is

$$n\lambda - \frac{me^2}{\hbar^2} = 0 \qquad n = 1, 2, 3 \dots$$

where $\lambda = \frac{\sqrt{-2mE}}{\hbar}$.

Substituting λ into the quantization-condition equation, and solving for the energy E, you get the following (keeping in mind that for each step, n is a quantum number, so $n = 1, 2, 3, \dots$):

$$\frac{n\sqrt{-2mE}}{\hbar} - \frac{me^2}{\hbar^2} = 0$$

$$\frac{n\sqrt{-2mE}}{\hbar} = \frac{me^2}{\hbar^2}$$

$$n^2 \frac{(-2mE)}{\hbar^2} = \frac{m^2 e^4}{\hbar^4}$$

$$-n^2 2E = \frac{me^4}{\hbar^2}$$

$$E = \frac{-me^4}{2n^2 \hbar^2}$$

Because E depends on the principal quantum number (see the previous section), you can rename this equation to E_n and make the quantum number values explicit:

$$E_n = \frac{-me^4}{2n^2 \hbar^2} \qquad n = 1, 2, 3 \dots$$

TIP

Physicists often write this result in terms of the Bohr radius — the orbital radius that Niels Bohr calculated for the electron in a hydrogen atom, r_0. The Bohr radius is $r_0 = \frac{\hbar^2}{me^2}$.

And in terms of r_0, here's what E_n equals:

$$E_n = \frac{-me^4}{2n^2 \hbar^2} = \frac{-e^2}{2r_0} \frac{1}{n^2} \qquad n = 1, 2, 3 \dots$$

The ground state, where $n = 1$, works out to be about E = –13.6 eV.

Notice that this energy is negative because the electron is in a bound state — you'd have to add energy to the electron to free it from the hydrogen atom. Here are the first and second excited states:

>> First excited state, $n = 2$: $E = -3.4$ eV

>> Second excited state, $n = 3$: $E = -1.5$ eV

REMEMBER

As I describe in Chapter 3, Niels Bohr used the idea of the quantum to find these excited state values in 1911, over a decade before the mathematics of quantum mechanics used in this chapter even existed. This feat rightfully earned him a Nobel Prize in Physics in 1922.

Crafting the form of the radial solution

In this section, you complete the calculation of the wave functions that begins in the earlier section, "Simplifying and splitting the hydrogen equation," and refer to the calculation of $R_{n\ell}(r)$ in the earlier section "Solving the radial Schrödinger equation." One way to think of this calculation process is that you repeatedly apply constraints, restructure, solve, and check boundary conditions and normalization until you get the final solution.

1. **Restructure the $R_{n\ell}(r)$ radial wave function equation.**

   From previous calculations, you know that $R_{n\ell}(r) = r^\ell f(r) e^{-\lambda r}$, where

   $f(r) = \sum_{k=0}^{N} a_k r^k$, and therefore, $R_{n\ell}(r) = r^\ell e^{-\lambda r} \sum_{k=0}^{N} a_k r^k$.

   But you have more work to do. The preceding equation comes from solving the radial Schrödinger equation:

   $$\frac{-\hbar^2}{2m}\frac{d^2}{dr^2} r R_{n\ell}(r) + \ell(\ell+1)\frac{\hbar^2}{2mr^2} r R_{n\ell}(r) - \frac{e^2}{r} r R_{n\ell}(r) = E_r r R_{n\ell}(r)$$

2. **Add a multiplicative constant, $A_{n\ell}$, to match that solution.**

   This constant depends on the principal quantum number n and the angular momentum quantum number ℓ, like this:

   $$R_{n\ell}(r) = A_{n\ell} r^\ell e^{-\lambda r} \sum_{k=0}^{N} a_k r^k$$

   You find $A_{n\ell}$ by normalizing $R_{n\ell}(r)$.

3. Solve $R_{n\ell}(r)$ by doing the math.

For example, try to find $R_{10}(r)$. In this case, $n = 1$ and $\ell = 0$. Then, because $N + \ell + 1 = n$, you have $N = n - \ell - 1$. So, $N = 0$ here. That makes $R_{n\ell}(r)$ look like this:

$$R_{10}(r) = A_{10}r^0 e^{-\lambda r} \sum_{k=0}^{0} a_k r^k$$

And the summation in this equation is equal to $\sum_{k=0}^{0} a_k r^k = a_0$, so

$$R_{10}(r) = A_{10}r^0 e^{-\lambda r} a_0$$

4. Restructure the $R_{10}(r)$ equation with substitutions and constraints.

You apply these assumptions and constraints:

- *You know that $r^0 = 1$, so $R_{10}(r) = A_{10}e^{-\lambda r} a_0$.*
- *You can rewrite $n\lambda - \dfrac{me^2}{\hbar^2} = 0$ (the quantum constraint on these*

  physical solutions) and $r_0 = \dfrac{\hbar^2}{me^2}$ (the equation for the Bohr radius)

  as $\lambda = \dfrac{me^2}{n\hbar^2} = \dfrac{1}{nr_0}$.
- *You can also write $R_{10}(r) = A_{10}e^{-\lambda r} a_0$ as*

  $$R_{10}(r) = A_{10}a_0 \exp\left(\frac{-r}{nr_0}\right), \text{ where } r_0 \text{ is the Bohr radius.}$$

Note: Because $R_{n\ell}$ is R_{10} in this case, $n = 1$ for this equation, but I left the n explicitly in the preceding equation just to make it clear how this would look in other energy states.

5. To find A_{10} and a_0, normalize the wave function $\psi_{100}(r, \theta, \phi)$ to 1.

You normalize by integrating $|\psi_{100}(r, \theta, \phi)|^2 d^3r$ over all space and setting the result equal to 1. This result represents the idea that the probability of this wave function existing somewhere in the universe is equal to 1. (See Chapters 4 and 8 for more about this probability.)

REMEMBER

Fortunately, the angular portion of the wave function is already normalized. By redefining $d^3r = r^2 \sin\theta\, dr\, d\theta\, d\phi$, and integrating the spherical harmonics, such as Y_{00}, over a complete sphere, $\int |Y_{00}|^2 \sin\theta d\theta d\phi$, you get 1. (You can find the calculation and normalization for spherical harmonics in Chapter 13 if you want to check it out.)

6. Normalize the radial part of the equation.

Substitute in for $R_{10}(r)$ to get

$$1 = \int_{0}^{+\infty} r^2 |R_{10}(r)|^2 dr$$

$$1 = A_{10}{}^2 a_0^2 \int_{0}^{+\infty} r^2 \exp\left(\frac{-2r}{nr_0}\right) dr$$

In this equation, remember that $n = 1$ and apply your math skills to solve these integrals. You ultimately end up with the following equation that gives you a relationship containing only A_{10} and a_0.

$$A_{10}a_0 = \frac{2}{\sqrt{r_0^3}}$$

7. **Finish the wave function solution with final substitutions.**

Though it takes a fair amount of work to get the preceding equation, it is actually a fairly simple result when you keep in mind that A_{10} is just there to normalize the result, and the equation $R_{10}(r)$ has the term $A_{10}a_0$ at the beginning. That's precisely what this solution shows, so you can just substitute it into $R_{10}(r)$ to get

$$R_{10}(r) = \frac{2}{\sqrt{r_0^3}}e^{\frac{-r}{r_0}}$$

You know that $\psi_{n\ell m}(r, \theta, \phi) = R_{n\ell}(r) Y_{\ell m}(\theta, \phi)$. Chapter 13 provides several of the normalized spherical harmonics, $Y_{\ell m}(\theta, \phi)$, including $Y_{00}(\theta, \phi) = \frac{1}{\sqrt{4\pi}}$

By substituting both R_{10} and Y_{00}, you get

$$\psi_{100}(r, \theta, \phi) = \frac{2}{\sqrt{r_0^3}}e^{\frac{-r}{r_0}}\frac{1}{\sqrt{4\pi}}$$

And that concludes the work to find the wave function for one of the simplest quantum states of the hydrogen atom.

Going General with the Hydrogen Wave Function

As you can no doubt tell from the preceding sections of this chapter, finding the wave function for a hydrogen atom is complicated. And remember, it is the simplest of the atoms! In general, here's what the wave function $\psi_{n\ell m}(r, \theta, \phi)$ looks like for hydrogen:

$$\psi_{n\ell m}(r, \theta, \phi) = \frac{\sqrt{\left(\frac{2}{nr_0}\right)^3}\sqrt{(n-\ell-1)!}}{\sqrt{2n(n-1)!}}e^{\frac{-r}{nr_0}}\left(\frac{2r}{nr_0}\right)^\ell L_{n-\ell-1}^{2\ell+1}\left(\frac{2r}{nr_0}\right)Y_{\ell m}(\theta, \phi)$$

where the term $L_{n-\ell-1}^{2\ell+1}\left(\frac{2r}{nr_0}\right)$ is called a *generalized Laguerre polynomial*.

This term is a form of solution to certain types of differential equations (including the type you've got here, fortunately). Here are the first few generalized Laguerre polynomials:

» $L_0^b(r) = 1$

» $L_1^b(r) = -r + b + 1$

» $L_2^b(r) = \dfrac{r^2}{2} - (b+2)r + \dfrac{(b+2)(b+1)}{2}$

» $L_3^b(r) = -\dfrac{r^3}{6} + \dfrac{(b+3)r^2}{2} - \dfrac{(b+2)(b+3)r}{2} + \dfrac{(b+1)(b+2)(b+3)}{6}$

In earlier sections of the chapter (see "Crafting the form of the radial solution"), you find that $\psi_{100}(r, \theta, \phi)$ looks like this:

$$\psi_{100}(r, \theta, \phi) = \frac{2}{\sqrt{r_0^3}} e^{\frac{-r}{r_0}} \frac{1}{\sqrt{4\pi}}$$

Here are some other hydrogen wave functions, which give you context for how elaborate they can get:

» $\psi_{20m}(r, \theta, \phi) = \dfrac{1}{\sqrt{2r_0^3}} \left(1 - \dfrac{r}{2r_0}\right) e^{\frac{r}{2r_0}} Y_{0m}(\theta, \phi)$

» $\psi_{21m}(r, \theta, \phi) = \dfrac{1}{\sqrt{6r_0^3}} \dfrac{r}{2r_0} e^{\frac{-r}{2r_0}} Y_{1m}(\theta, \phi)$

» $\psi_{300}(r, \theta, \phi) = \dfrac{2}{3\sqrt{3r_0^3}} e^{\frac{-r}{3r_0}} \left[1 - \left(\dfrac{2r}{3r_0}\right) + \left(\dfrac{2r^2}{27r_0^2}\right)\right] Y_{00}(\theta, \phi)$

» $\psi_{31m}(r, \theta, \phi) = \dfrac{8}{9\sqrt{6r_0^3}} e^{\frac{-r}{3r_0}} \dfrac{r}{3r_0} \left[1 - \left(\dfrac{r}{6r_0}\right)\right] Y_{1m}(\theta, \phi)$

» $\psi_{32m}(r, \theta, \phi) = \dfrac{4}{9\sqrt{30r_0^3}} \dfrac{r^2}{9r_0^2} e^{\frac{-r}{3r_0}} Y_{2m}(\theta, \phi)$

Note that $\psi_{n\ell m}(r, \theta, \phi)$ behaves like r^ℓ for small r and therefore goes to zero. And for large r, $\psi_{n\ell m}(r, \theta, \phi)$ decays exponentially to zero. So you've solved the problem from the earlier section, "Solving the radial Schrödinger equation," of the wave function diverging as r becomes large. Modifying that summation so that f(r) didn't become infinite turned out to be useful. Not bad.

You can see the radial wave functions $R_{10}(r)$, $R_{20}(r)$, and $R_{21}(r)$ in Figure 14-2.

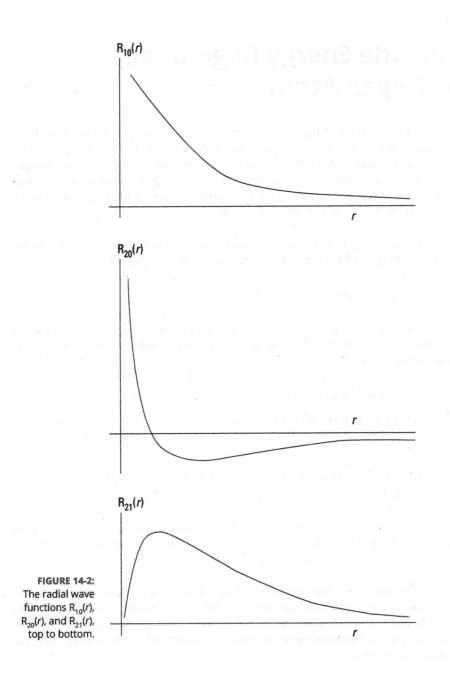

FIGURE 14-2:
The radial wave
functions $R_{10}(r)$,
$R_{20}(r)$, and $R_{21}(r)$,
top to bottom.

Calculating the Energy Degeneracy of the Hydrogen Atom

Each quantum state of the hydrogen atom ($\psi_{n\ell m}[r, \theta, \phi]$) is specified with three quantum numbers: n (the principal quantum number), ℓ (the angular momentum quantum number of the electron), and m (the z component of the electron's angular momentum). You might wonder how many of these states have the same energy. In other words, what's the energy degeneracy of the hydrogen atom in terms of the quantum numbers n, ℓ, and m?

The actual energy is just dependent on n, as found in the section "Finding the allowed energies of the hydrogen atom," earlier in this chapter:

$$E_n = \frac{-me^4}{2n^2\hbar^2} \quad n = 1, 2, 3 \ldots$$

From this equation, you can see that E is independent of ℓ and m. So how many states, $|n, l, m\rangle$, have the same energy for a particular value of n? You have these circumstances:

» **For a particular value of n,** ℓ can range from zero to $n-1$.

» **For any particular value of ℓ,** you can have m values of $-\ell, -\ell + 1, \ldots, 0, \ldots, \ell - 1, \ell$. That means that $(2\ell + 1)$ possible m states exist for a particular value of ℓ.

» **For each value of n,** you have n^2 states, all of which would have the same energy level.

So, the degeneracy of the energy levels of the hydrogen atom is n^2. For example, the ground state, $n = 1$, has *degeneracy* $= n^2 = 1$. This makes sense because ℓ, and therefore m, can only equal zero for this state.

For $n = 2$, you have a degeneracy of 4. Again, this value is small enough that you can easily think through all the cases. You have $\ell = 0$ and $m = 0$ as one state. You also have $\ell = 1$, which allows three states, where $m = -1$, $m = 0$, and $m = 1$. The following wave functions represent those four states, all of which would have the same energy values:

» $\psi_{200}(r, \theta, \phi)$

» $\psi_{21-1}(r, \theta, \phi)$

» $\psi_{210}(r, \theta, \phi)$

» $\psi_{211}(r, \theta, \phi)$

Quantum states: Adding a little spin

You may be asking yourself, "What about the spin of the electron?" Right you are! The spin of the electron does provide additional quantum states. So far in this section, you can treat the wave function of the hydrogen atom as a product of radial and angular parts:

$$\psi_{n\ell m}(r, \theta, \phi) = R_{n\ell}(r)Y_{\ell m}(\theta, \phi)$$

Now you can add a spin part, corresponding to the spin of the electron, where s is the spin of the electron and m_s is the z component of the spin:

$$|s, m_s\rangle$$

The spin part of the equation can take the following values:

- » $\left|\frac{1}{2}, \frac{1}{2}\right\rangle$
- » $\left|\frac{1}{2}, -\frac{1}{2}\right\rangle$

Wave functions for spin states

When you take spin into account, $\psi_{n\ell m}(r, \theta, \phi)$ becomes $\psi_{n\ell m m_s}(r, \theta, \phi)$:

$$\psi_{n\ell m m_s}(r, \theta, \phi) = R_{n\ell}(r)Y_{\ell m}(\theta, \phi)|s, m_s\rangle$$

And this wave function can take two forms, depending on m_s, like this:

- » $\psi_{n\ell m-\frac{1}{2}}(r, s, \theta, \phi) = R_{n\ell}(r)Y_{\ell m}(\theta, \phi)\left|\frac{1}{2}, \frac{1}{2}\right\rangle$
- » $\psi_{n\ell m-\frac{1}{2}}(r, s, \theta, \phi) = R_{n\ell}(r)Y_{\ell m}(\theta, \phi)\left|\frac{1}{2}, -\frac{1}{2}\right\rangle$

Spin's effect on degeneracy

What does adding spin to the wave functions do to the energy degeneracy? If you include the spin of the electron, two spin states exist for every state $|n, \ell, m\rangle$. So, if you include the electron's spin, the energy degeneracy of the hydrogen atom is $2n^2$.

In fact, you can even add the spin of the proton to the wave function (although people don't usually do that, because the proton's spin interacts only weakly with magnetic fields applied to the hydrogen atom). In that case, you have a wave function that looks like the following:

$$\psi_{n\ell m m_{se} m_{sp}}(r, \theta, \phi) = R_{n\ell}(r) Y_{\ell m}(\theta, \phi) |s_e, m_{se}\rangle |s_p, m_{sp}\rangle$$

where s_e is the spin of the electron, m_{se} is the z component of the electron's spin, s_p is the spin of the proton, and m_{sp} is the z component of the proton's spin. The electron's spin, m_{se}, has two states, up or down, and so does the proton's spin, m_{sp}.

You can add together these two sets of two independent states. The degeneracy when including the proton's spin (as well as the electron's) results in a factor of four for each $|n, \ell, m\rangle$. And the degeneracy is now $4n^2$!

On the lines: Getting the orbitals

When you study heated hydrogen in spectroscopy, you get a spectrum consisting of various lines, named the s (for *sharp*), p (for *principal*), d (for *diffuse*), and f (for *fundamental*) lines. Other, unnamed lines — the g, h, and so on — are present as well.

The s, p, d, f, and the rest of the lines turn out to correspond to different angular momentum states of the electron, called *orbitals*. The s state corresponds to $\ell = 0$; the p state, to $\ell = 1$; the d state, to $\ell = 2$; the f state, to $\ell = 3$; and so on. Each of these angular momentum states has a differently shaped electron cloud around the proton — that is, a different orbital.

Three quantum numbers — n, ℓ, and m — determine orbitals. Figure 14-3 shows the electron clouds for the $|1, 0, 0\rangle$ state (1s, with m = 0); the $|3, 2, 1\rangle$ state (4f, with m = 2); and the $|2, 1, 1\rangle$ state (2p, with m = 1).

FIGURE 14-3:
The electron clouds for the $|1, 0, 0\rangle$, $|3, 2, 1\rangle$, and $|2, 1, 1\rangle$ states.

The $|1, 0, 0\rangle$ state   The $|3, 2, 1\rangle$ state   The $|2, 1, 1\rangle$ state

Hunting the Elusive Electron

You may be curious where the electron in a hydrogen atom is at any one time. In other words, how far is the electron from the proton? You can find the expectation value of r, that is, $\langle r \rangle$, to tell you. If the wave function is $\psi_{n\ell m}(r, \theta, \phi)$, then the following expression represents the probability that the electron will be found in the spatial element d^3r:

$$|\psi_{n\ell m}(r, \theta, \phi)|^2 d^3r$$

In spherical coordinates, $d^3r = r^2 \sin\theta\, dr\, d\theta\, d\phi$, so you can write $|\psi_{n\ell m}(r, \theta, \phi)|^2 d^3r$ as

$$|\psi_{n\ell m}(r, \theta, \phi)|^2 r^2 \sin\theta\, dr\, d\theta\, d\phi$$

The probability that the electron is in a spherical shell of radius r to $r + dr$ is therefore

$$\int_0^{2\pi}\int_0^{\pi} |\psi_{n\ell m}(r, \theta, \phi)|^2 r^2 \sin\theta\, d\theta\, d\phi\, dr$$

And because $\psi_{n\ell m}(r, \theta, \phi) = R_{n\ell}(r)Y_{\ell m}(\theta, \phi)$, this equation becomes the following:

$$\int_0^{2\pi}\int_0^{\pi} |R_{n\ell}(r)Y_{\ell m}(\theta, \phi)|^2 r^2 \sin\theta\, d\theta\, d\phi\, dr$$

The preceding integral is equal to

$$|R_{n\ell}(r)|^2 r^2 dr \int_0^{\pi} |Y_{\ell m}(\theta, \phi)|^2 \sin\theta\, d\theta \int_0^{2\pi} d\phi$$

Spherical harmonics Y are normalized (see Chapter 13), so those integrals are equal to 1 and the preceding equation just becomes

$$|R_{n\ell}(r)|^2 r^2 dr$$

That's the probability that the electron is inside the spherical shell from r to $r + dr$. The expectation value of r is

$$\langle r \rangle = \int_0^{\infty} r |R_{n\ell}(r)|^2 r^2 dr$$

which is

$$\langle r \rangle = \int_0^{\infty} r^3 |R_{n\ell}(r)|^2 dr$$

This is where things get more complex, because $R_{n\ell}(r)$ involves the Laguerre polynomials. But after a lot of math, here's what you get:

$$\langle r \rangle = \int_0^\infty r^3 |R_{n\ell}(r)|^2 \, dr = \left[3n^2 - \ell(\ell+1) \right] \frac{r_0}{2}$$

where r_0 is the Bohr radius: $r_0 = \dfrac{\hbar^2}{me^2}$. The Bohr radius is about 5.29×10^{-11} meters, so the expectation value of the electron's distance from the proton is

$$\langle r \rangle = \left[3n^2 - \ell(\ell+1) \right] \left(2.65 \times 10^{-11} \right) \text{ meters.}$$

So, for example, in the 1s state ($|1, 0, 0\rangle$), the expectation value of r is equal to

$$\langle r \rangle_{1s} = 3\left(2.65 \times 10^{-11} \right) = 7.95 \times 10^{-11} \text{ meters.}$$

And in the 4p state ($|4, 1, m\rangle$),

$$\langle r \rangle_{4p} = 46\left(2.65 \times 10^{-11} \right) = 1.22 \times 10^{-9} \text{ meters.}$$

IN THIS CHAPTER

Chapter **15**

Handling Many Particles and Group Dynamics

E arlier in the book, most of the chapters that dove into the mathematics of quantum physics (see Chapters 7 through 14) focused on the behavior of a single particle. This focus was for a good reason: Quantum physics equations involving multiple particles become complex very quickly, and they are usually so complex that finding an exact solution to the equation is impossible. Because the unsolvable situation is not exactly the best place to find out how something works, in this chapter, I focus on problems involving simple systems where a precise solution can be calculated.

If you have grasped the basic concepts from previous chapters, you can dig into exploring more complex systems. Even without finding exact wave functions, you can still discover a surprising amount about multi-particle systems. For example, you can derive the Pauli exclusion principle, which says, among other things, that no two electrons can be in the exact same quantum state. But many multi-particle systems are so complex that the work needed to describe a new situation usually involves teams of researchers collaborating as part of major research projects and doctoral dissertations. This chapter only skims the surface. (You can find and work through problems in greater detail in *Quantum Physics Workbook For Dummies* [Wiley], if you're so inclined.)

This chapter starts with an introduction to many-particle systems and goes on to discuss identical particles, symmetry (and antisymmetry), and electron shells. I then help you explore perturbation theory, which gives you the ability to combine concepts from different types of quantum behavior together. And finally, I give you a (mathematical) glimpse inside particle accelerators, so you can see how particle physicists use scattering theory to describe the outcomes of particle collisions.

Note: In this chapter, you apply the mathematics introduced in Chapters 7 through 14, but you don't actually solve any equations. This chapter is intended as something of a springboard for any work you'd do with multiple particles. If you feel completely comfortable with the math in the previous chapters and want to dive more deeply into systems with multiple particles, you need a deeper understanding of multivariable calculus. And it might be a good idea to begin looking into doctoral programs in your local university's physics department.

Many-Particle Systems, Generally Speaking

In Chapter 14, I explore some of the basic quantum physics mathematics around the hydrogen atom. These are the simplest atoms, because hydrogen involves only a single proton and (for the versions I examine, at least) a single electron. Even so, those equations in Chapter 14 are pretty complex and require all the tools explored in Parts 3 and 4 to solve them. So, what happens when you have an even more elaborate system, where you are dealing with more particles?

Figure 15-1 depicts a multi-particle system that identifies a number of particles by their positions in a three-dimensional space (ignore spin for the moment). This section explains how to describe that system in quantum physics terms.

Considering wave functions and Hamiltonians

Begin by working with the wave function. The state of a system with many *particles* (refer to Figure 15-1) is given by $\psi(r_1, r_2, r_3, \ldots)$. And here's the probability that particle 1 is in d^3r_1, particle 2 is in d^3r_2, particle 3 is in d^3r_3, and so on:

$$|\psi(r_1, r_2, r_3, \ldots)|^2 \, d^3 r_1 d^3 r_2 d^3 r_3 \ldots$$

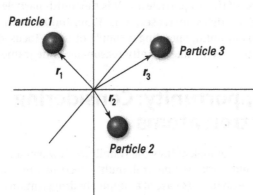

Particle 1

Particle 3

r_1

r_3

r_2

Particle 2

The normalization of $\psi(r_1, r_2, r_3, \ldots)$ demands that there is a probability that these particles exist somewhere, and integrating the preceding function across all of space must equal a probability of 1. (See Chapter 8 for more on normalization.)

$$\int_{-\infty}^{+\infty} |\psi(r_1, r_2, r_3, \ldots)|^2 \, d^3r_1 d^3r_2 d^3r_3 \ldots = 1$$

What about the Hamiltonian operator, H, which gives you the energy states? The Hamiltonian is a very useful thing to have in quantum physics (as discussed in Chapter 7). What is H, where $H\psi(r_1, r_2, r_3, \ldots) = E\psi(r_1, r_2, r_3, \ldots)$? When you're dealing with a single particle, you can write the Hamiltonian operator as

$$\frac{p^2}{2m}\psi(r) + V(r)\psi(r) = E\psi(r)$$

But in a many-particle system, the Hamiltonian must represent the total energy of all particles, not just one. The total energy of the system is the sum of the energy of all the particles (omitting spin for the moment), so here's how you can generalize the Hamiltonian for multi-particle systems:

$$H\psi(r_1, r_2, r_3, \ldots) = \sum_{i=1}^{N} \frac{p_i^2}{2m_i} \psi(r_1, r_2, r_3, \ldots) + V(r_1, r_2, r_3, \ldots)\psi(r_1, r_2, r_3, \ldots)$$

This, in turn, equals the following:

$$H\psi(r_1, r_2, r_3, \ldots) = \sum_{i=1}^{N} \frac{p_i^2}{2m_i} \psi(r_1, r_2, r_3, \ldots) + V(r_1, r_2, r_3, \ldots)\psi(r_1, r_2, r_3, \ldots)$$

$$= \sum_{i=1}^{N} \frac{-\hbar^2}{2m_i} \nabla_i^2 \psi(r_1, r_2, r_3, \ldots) + V(r_1, r_2, r_3, \ldots)\psi(r_1, r_2, r_3, \ldots)$$

Here, m_i is the mass of the ith particle and V is the multi-particle potential. Again, as the number of particles increases, this Hamiltonian becomes increasingly unwieldy. With this in mind, spending a couple of years focused on just a single multi-particle system seems like a pretty reasonable time frame.

A Nobel opportunity: Considering multi-electron atoms

This section takes a brief look at how the Hamiltonian wave function (see the preceding section) would work for a neutral, multi-electron atom (which is a significant step up from the Chapter 14 case of a simpler hydrogen atom). A multi-electron atom, as depicted in Figure 15-2, is a common multi-particle system that quantum physics considers. Here, **R** is the coordinate of the nucleus (relative to the center of mass), r_1 is the coordinate of the first electron (relative to the center of mass), r_2 is the coordinate of the second electron, and so on.

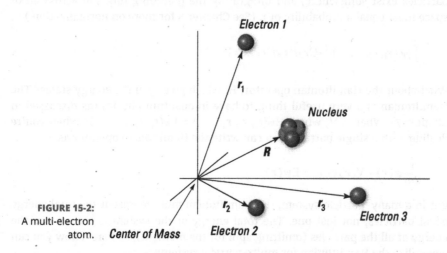

FIGURE 15-2:
A multi-electron atom.

If you have Z electrons, the wave function looks like $\psi(r_1, r_2, \ldots, r_Z, \mathbf{R})$, while the kinetic energy of the electrons and the nucleus looks like this:

$$KE = \sum_{i=1}^{Z} \frac{-\hbar^2}{2m_i} \nabla_i^2 \psi(r_1, r_2, \ldots r_Z, \mathbf{R}) - \frac{\hbar^2}{2M} \nabla_R^2 \psi(r_1, r_2, \ldots r_Z, \mathbf{R})$$

And the potential energy of the system looks like this:

$$PE = -\sum_{i=1}^{Z} \frac{Ze^2}{|r_i - \mathbf{R}|} \psi(r_1, r_2, \ldots r_Z, \mathbf{R}) + \sum_{i>j} \frac{e^2}{|r_i - r_j|} \psi(r_1, r_2, \ldots r_Z, \mathbf{R})$$

So, adding the two preceding equations, here's what you get for the total energy (E = KE + PE) of a multi-particle atom:

$$E\psi(r_1, r_2, \ldots r_z, R) = \sum_{i=1}^{Z} \frac{-\hbar^2}{2m_i} \nabla_i^2 \psi(r_1, r_2, \ldots r_z, R) - \frac{\hbar^2}{2M} \nabla_R^2 \psi(r_1, r_2, \ldots r_z, R)$$

$$- \sum_{i=1}^{Z} \frac{Ze^2}{|r_i - R|} \psi(r_1, r_2, \ldots r_z, R) + \sum_{i>j}^{Z} \frac{e^2}{|r_i - r_j|} \psi(r_1, r_2, \ldots r_z, R)$$

Okay, now that looks like a proper mess. Want to win the Nobel Prize in Physics? Just come up with the general solution to the preceding equation. Whenever you have a multi-particle system in which the particles interact with each other, you can't split this equation into a system of N independent equations.

REMEMBER

In cases where the N particles of a multi-particle system don't interact with each other, you can disconnect the Schrödinger equation into a set of N independent equations, and solutions may be possible. But when the particles interact and the Schrödinger equation depends on those interactions, you can't solve that equation for any significant number of particles.

Looking at a super-powerful tool: Interchange symmetry

Even though finding general solutions for equations like the one for the total energy of a multi-particle atom (in the preceding section) is impossible, you can still see what happens when you exchange particles with each other — and the results are very revealing. This section covers the idea of interchange symmetry.

Order matters: Swapping particles with the exchange operator

You can determine what happens to the wave function when you swap two particles. Whether the wave function is symmetric under such operations offers insight into whether two particles can occupy the same quantum state. This section discusses swapping particles and looking at symmetric and antisymmetric functions.

Take a look at the general wave function for N particles:

$$\psi(r_1, r_2, \ldots, r_i, \ldots, r_j, \ldots, r_N)$$

SYMMETRIES ABOUND

In this chapter, I talk about symmetry in terms of the location coordinate, r, to keep things simple, but you can also consider other quantities, such as spin, velocity, and so on. That wouldn't make this discussion any different because you can wrap all of a particle's quantum measurements — location, velocity, speed, and so on — into a single quantum state, which you can call ξ (the Greek letter xi). Doing so would make the general wave function for N particles into this: $\psi(\xi_1, \xi_2, \ldots, \xi_i, \ldots, \xi_j, \ldots, \xi_N)$. But remember, this section just considers the wave function $\psi(r_1, r_2, \ldots, r_i, \ldots, r_j, \ldots, r_N)$ to keep things simple.

In physics, symmetry represents ways you can transform a system without changing the underlying reality of what you're talking about. In his 1972 article, "More Is Different," published in *Science* (Vol. 177, No. 4047), American theoretical physicist Philip W. Anderson said, "By *symmetry* we mean the existence of different viewpoints from which the system appears the same. It is only slightly overstating the case to say that physics is the study of symmetry." Much of advanced theoretical and particle physics is explicitly expressed in terms of the symmetries that are present, and the symmetries that aren't present. Symmetries that aren't present, or which do not hold, are referred to in physics as "broken" symmetries. These broken symmetries can have profound implications, and several Nobel Prizes in Physics have been awarded for analyzing broken symmetries in physics.

For example, our universe appears to consist of much more *baryonic matter* (the technical term for regular matter) than antimatter. If the amount of baryonic matter and antimatter had been perfectly balanced at the creation of the universe, you could expect that these opposing matters would have annihilated each other, and no matter would be left. The reason that didn't happen is because a *baryon asymmetry* existed in the initial distribution, so the universe had more baryonic matter than antimatter — which resulted in a pretty important consequence of this physical asymmetry!

Now imagine that you have an exchange operator, P_{ij}, that exchanges particles i and j. In other words, it is an operator that takes the r_i term and swaps it with the r_j term. (That is literally all this operator does.) By using the exchange operator, you can identify two results right away in this situation:

>> Swapping the i term with the j term is identical to swapping the j term with the i term, so $P_{ij} = P_{ji}$.

>> Using P_{ij} twice swaps two terms and then immediately swaps them back, so P_{ij}^2 results in the original function. This means $P_{ij}^2 = 1$.

Classifying symmetric and antisymmetric wave functions

$P_{ij}^2 = 1$ (see the preceding section), so note that if a wave function is an eigenfunction of P_{ij}, then the possible eigenvalues are 1 and -1. That is, for $\psi(r_1, r_2, \ldots, r_i, \ldots, r_j, \ldots, r_N)$ an eigenfunction of P_{ij} looks like this:

$$P_{ij}\psi(r_1, r_2, \ldots, r_i, \ldots, r_j, \ldots, r_N) = \psi(r_1, r_2, \ldots r_i, \ldots, r_j, \ldots, r_N) \text{ or}$$

$$-\psi(r_1, r_2, \ldots, r_i, \ldots, r_j, \ldots, r_N)$$

You have two kinds of eigenfunctions of the exchange operator:

>> Symmetric eigenfunctions:

• $P_{ij}\psi_s(r_1, r_2, \ldots, r_i, \ldots, r_j, \ldots, r_N) = \psi_s(r_1, r_2, \ldots, r_i, \ldots, r_j, \ldots, r_N)$

>> Antisymmetric eigenfunctions:

• $P_{ij}\psi_a(r_1, r_2, \ldots, r_i, \ldots, r_j, \ldots, r_N) = -\psi_a(r_1, r_2, \ldots, r_i, \ldots, r_j, \ldots, r_N)$

If the wave function doesn't fit one of these formulas, then that wouldn't be an eigenfunction of the operator P_{ij}. (See Chapter 7 for an introduction to eigenfunctions.) In a situation where all the particles are identical — like an atom with multiple electrons — this symmetry seems fairly straightforward. Things do get a little bit more complicated when the particles aren't identical.

Floating cars: Tackling systems of many distinguishable particles

The preceding section discusses the idea of swapping like particles; now, you look at systems of particles that you can distinguish — that is, systems of identifiably different particles. As you will see in this section, you can decouple such systems into linearly independent equations.

Suppose you have a system of many different types of cars floating around in space. You can distinguish all those cars because they're all different — they have different masses, for one thing.

Now say that each car interacts with its own potential — that is, the potential that any one car sees doesn't depend on any other car. That means that the potential for all cars is just the sum of the individual potentials each car sees, which looks like this, assuming you have N cars:

$$PE = V(r_1, r_2, \ldots, r_N) = \sum_{i=1}^{N} V(r_i)$$

Being able to cut the potential energy up into a sum of independent terms like this makes life a lot easier. Here's what the Hamiltonian looks like:

$$H\psi(r_1, r_2, ..., r_N) = \sum_{i=1}^{N}\left[\frac{-\hbar^2}{2m_i}\nabla_i^2 + V_i(r_i)\right]\psi(r_1, r_2, ..., r_N)$$

This is much simpler than the Hamiltonian for a hydrogen atom from earlier in the chapter. Note that you can separate the previous equation for the potential of all cars into N different equations:

$$\frac{-\hbar^2}{2m_i}\nabla_i^2\psi_i(r_i) + V_i(r_i)\psi_i(r_i) = E_i\psi_i(r_i)$$

And the total energy is just the sum of the energies of the individual cars:

$$E = \sum_{i=1}^{N}E_i$$

And the wave function is just the product of the individual wave functions:

$$\psi_{n_1, n_2, ..., n_N}(r_1, r_2, ..., r_N) = \prod_{i=1}^{N}\psi_{n_i}(r_i)$$

where the Π symbol is like the Σ symbol, except that it stands for a product of terms, not a summation, and n_i refers to all the quantum numbers of the ith particle.

REMEMBER

When the particles in the system you're working with are distinguishable and subject to independent potentials, the problem of handling many of them becomes simpler. You can break the system up into N independent one-particle systems. The total energy is just the sum of the individual energy of each particle. The Schrödinger equation breaks down into N different equations. And the wave function ends up just being the product of the wave functions of the N different particles.

Juggling many identical particles

When the particles in a multi-particle system are all indistinguishable, that's when the real adventure begins. When you can't tell the particles apart, how can you tell which one's where? This section explains what happens.

Losing identity

Say that you have a bunch of pool balls (as in billiards, not as in balls you use in the swimming pool) and you want to look at them classically. You can paint each pool ball differently, and then, even as they hurtle around the pool table, you're able to distinguish them — 7-ball in the corner pocket, and that sort of thing. Classically, identical particles retain their individuality. You can still tell them apart.

The same isn't true quantum mechanically because you can't locate particles with absolute precision. So if you were to have a bunch of electrons, you'd quickly lose track of which one was which — you can't paint them, as you can pool balls. Nothing about two different electrons allows you to really distinguish one of them from the other!

For example, look at the scenario in Figure 15-3, which depicts two electrons colliding and bouncing apart. In this case, keeping track of the two electrons would seem to be easy.

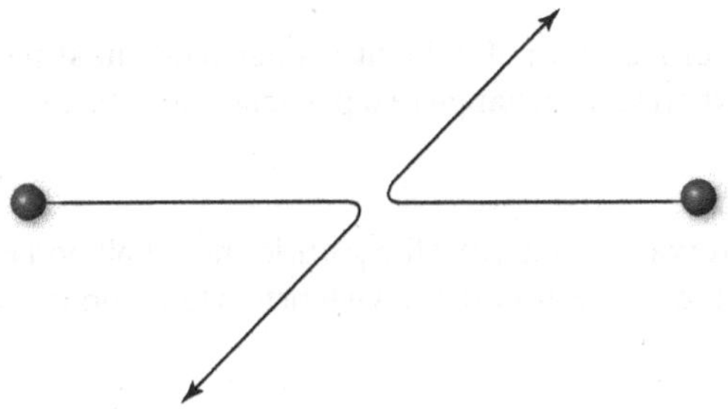

FIGURE 15-3:
An electron colliding with another electron.

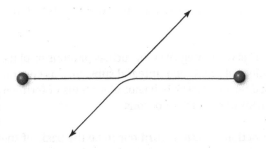

But now look at the scenario in Figure 15-4. The same two electrons could've bounced like the depiction in that figure and not like the bounce shown in Figure 15-3 — and you'd never know it.

FIGURE 15-4:
An electron colliding with another electron.

So which electron is which? From the experimenter's point of view, you can't tell. You can place detectors to catch the electrons, but you can't determine which of the incoming electrons ended up in which detector because of the two possible scenarios shown in Figures 15-3 and 15-4.

Quantum mechanically, identical particles don't retain their individuality in terms of any measurable, observable quantity. You lose the individuality of identical particles as soon as you mix them with similar particles. This idea holds true for any N-particle system. As soon as you let N identical particles interact, you can't say which exact one is at r_1 or r_2 or r_3 or r_4 and so on.

Symmetry and antisymmetry

In practical terms, the loss of individuality among identical particles means that the *probability density* (the probability of a continuous random variable lying between a specific range of values) remains unchanged when you exchange particles. For example, if you were to exchange electron 10,281 with electron 59,830, you'd still have the same probability that an electron would occupy $d^3r_{10,281}$ and $d^3r_{59,830}$.

The wave function of a system of N identical particles must be either symmetric or antisymmetric when you exchange two particles. Spin turns out to be the deciding factor:

>> **Antisymmetric wave function:** If the particles have half-odd-integral spin ($\frac{1}{2}$, $\frac{3}{2}$, and so on), then this is how the wave function looks under exchange of particles:

- $\psi(r_1 s_1, r_2 s_2, \ldots, r_i s_i, \ldots, r_j s_j, \ldots, r_N s_N) = -\psi(r_1 s_1, r_2 s_2, \ldots, r_j s_j, \ldots, r_i s_i, \ldots, r_N s_N)$

>> **Symmetric wave function:** If the particles have integral spin (0, 1, and so on), this is how the wave function looks under exchange of particles:

- $\psi(r_1 s_1, r_2 s_2, \ldots, r_i s_i, \ldots, r_j s_j, \ldots, r_N s_N) = \psi(r_1 s_1, r_2 s_2, \ldots, r_j s_j, \ldots, r_i s_i, \ldots, r_N s_N)$

Having symmetric or antisymmetric wave functions leads to some different physical behaviors.

In particular, particles with integral spin, such as photons or pi mesons, are called bosons. And particles with half-odd-integral spin, such as electrons, protons, and neutrons, are called fermions. The behavior of systems of fermions is very different from the behavior of systems of bosons.

And these wave function equations turn out to be the basis of the *symmetrization postulate*, which says that in systems of N identical particles, only states that are symmetric or antisymmetric exist — and it says that states of mixed symmetry don't exist.

The symmetrization postulate also says, as observed from nature, that

>> **Fermions** have antisymmetric states under the interchange of two particles.

>> **Bosons** have symmetric states under the interchange of two particles.

Chapters 4 and 5 cover fermions and bosons in greater detail. In this context, the key thing to know is that the wave function of N fermions is completely antisymmetric, and the wave function of N bosons is completely symmetric.

Exchange degeneracy: The steady Hamiltonian

The Hamiltonian doesn't vary under the exchange of two identical particles. In other words, the Hamiltonian is invariant, no matter how many identical particles you exchange. That's called exchange degeneracy.

Incidentally, the exchange operator, P_{ij}, (introduced in the earlier section "Order matters: Swapping particles with the exchange operator") is an invariant of the motion because it commutes with the Hamiltonian:

$[H, P_{ij}] = 0$

Building symmetric and antisymmetric wave functions

Many of the wave functions that are solutions to physical setups like the square well aren't inherently symmetric or antisymmetric; they're simply asymmetric. In other words, they have no definite symmetry. So how do you end up with symmetric or antisymmetric wave functions?

The answer is that you have to create them yourself, and you do that by adding together asymmetric wave functions. For example, suppose that you have an asymmetric wave function of two particles, $\psi(r_1 s_1, r_2 s_2)$.

To create a symmetric wave function, add together $\psi(r_1 s_1, r_2 s_2)$ and the version where the two particles are swapped, $\psi(r_2 s_2, r_1 s_1)$. Assuming that $\psi(r_1 s_1, r_2 s_2)$ and $\psi(r_2 s_2, r_1 s_1)$ are normalized, you can create a symmetric wave function using these two wave functions — just by adding them:

$$\psi_s(r_1 s_1, r_2 s_2) = \frac{1}{\sqrt{2}} [\psi(r_1 s_1, r_2 s_2) + \psi(r_2 s_2, r_1 s_1)]$$

You can make an antisymmetric wave function by subtracting the two wave functions:

$$\psi_a(r_1s_1, r_2s_2) = \frac{1}{\sqrt{2}}[\psi(r_1s_1, r_2s_2) - \psi(r_2s_2, r_1s_1)]$$

This process rapidly becomes more complex when you add more particles because you have to interchange all the particles. For example, if you were writing the wave function for a three-particle system, then you'd have six distinct wave functions that you would be adding together (for a symmetric wave function) or adding and subtracting (for the asymmetric wave function).

In theory, you can create symmetric and antisymmetric wave functions for any system of N particles by combining the wave functions of swapped particles. In practice, creating the wave functions becomes a nightmare pretty quickly.

Working with Identical Noninteracting Particles

Working with identical noninteracting particles makes life easier because you can treat the equations individually instead of combining them into one big mess. Suppose that you have a system of N identical particles, each of which experiences the same potential. You can separate the Schrödinger equation into N identical single-particle equations. This idea is similar to what I describe in the earlier section "Floating Cars: Tackling systems of many Distinguishable Particles."

In that section, you consider the wave function of a system of N *distinguishable* particles and come up with the product of all the individual wave functions. But that equation doesn't work with identical particles because

>> You can't say that particle 1 is in state $\psi_1(r_1)$, particle 2 is in state $\psi_2(r_2)$, and so on — they're identical particles here, not distinguishable particles as before.

>> It has no inherent symmetry — and systems of N identical particles must have a definite symmetry. So, instead of simply multiplying the wave functions, you have to be a little more careful.

Wave functions of two-particle systems

How do you create symmetric and antisymmetric wave functions for a two-particle system? Start with the single-particle wave functions (see the earlier section "Building symmetric and antisymmetric wave functions"):

$$\gg \quad \psi_s(r_1s_1, r_2s_2) = \frac{1}{\sqrt{2}}[\psi(r_1s_1, r_2s_2) + \psi(r_2s_2, r_1s_1)]$$

$$\gg \quad \psi_a(r_1s_1, r_2s_2) = \frac{1}{\sqrt{2}}[\psi(r_1s_1, r_2s_2) - \psi(r_2s_2, r_1s_1)]$$

By analogy, here's the symmetric wave function, this time made up of two single-particle wave functions:

$$\psi_s(r_1s_1, r_2s_2) = \frac{1}{\sqrt{2}}[\psi_{n_1}(r_1s_1)\psi_{n_2}(r_2s_2) + \psi_{n_1}(r_2s_2)\psi_{n_2}(r_1s_1)]$$

And here's the antisymmetric wave function, made up of the two single-particle wave functions:

$$\psi_a(r_1s_1, r_2s_2) = \frac{1}{\sqrt{2}}[\psi_{n_1}(r_1s_1)\psi_{n_2}(r_2s_2) - \psi_{n_1}(r_2s_2)\psi_{n_2}(r_1s_1)]$$

where n_i stands for all the quantum numbers of the ith particle.

Note in particular that $\psi_a(r_1s_1, r_2s_2) = 0$ when $n_1 = n_2$; in other words, the antisymmetric wave function vanishes when the two particles have the same set of quantum numbers — that is, when they're in the same quantum state. That idea has important physical ramifications (see the later section "It's not come one, come all: The Pauli exclusion principle").

Here are three other ways to express the wave function $\psi_s(r_1s_1, r_2s_2)$:

$\gg$ **Where P is the permutation operator** (which takes the permutation of its argument), you have

$$\psi_s(r_1s_1, r_2s_2) = \frac{1}{\sqrt{2!}}\sum_P P\psi_{n_1}(r_1s_1)\psi_{n_2}(r_2s_2)$$

$\gg$ **Where the term $(-1)^P$ is 1 for even permutations** (where you exchange both r_1s_1 and r_2s_2 and also n_1 and n_2) and -1 for odd permutations (where you exchange r_1s_1 and r_2s_2 but not n_1 and n_2; or you exchange n_1 and n_2 but not r_1s_1 and r_2s_2), you have

$$\psi_a(r_1s_1, r_2s_2) = \frac{1}{\sqrt{2!}}\sum_P (-1)^P P\psi_{n_1}(r_1s_1)\psi_{n_2}(r_2s_2)$$

$\gg$ **In determinant form**, you have

$$\psi_a(r_1s_1, r_2s_2) = \frac{1}{\sqrt{2!}} \det \begin{vmatrix} \psi_{n_1}(r_1s_1) & \psi_{n_1}(r_2s_2) \\ \psi_{n_2}(r_1s_1) & \psi_{n_2}(r_2s_2) \end{vmatrix}$$

Note that this determinant is zero if $n_1 = n_2$.

Wave functions of three-or-more-particle systems

Now you get to put together the wave function of a system of three particles from single-particle wave functions. The symmetric and antisymmetric wave functions (respectively) look like this:

$$\psi_s(r_1s_1, r_2s_2, r_3s_3) = \frac{1}{\sqrt{3!}} \sum_P P\psi_{n_1}(r_1s_1)\psi_{n_2}(r_2s_2)\psi_{n_3}(r_3s_3)$$

$$\psi_a(r_1s_1, r_2s_2, r_3s_3) = \frac{1}{\sqrt{3!}} \sum_P (-1)^P \psi_{n_1}(r_1s_1)\psi_{n_2}(r_2s_2)\psi_{n_3}(r_3s_3)$$

How about generalizing this to systems of N particles? If you have a system of N particles, the symmetric and antisymmetric wave functions (respectively) look like this:

$$\psi_s(r_1s_1, r_2s_2, ..., r_Ns_N) = \frac{1}{\sqrt{N!}} \sum_P P\psi_{n_1}(r_1s_1)\psi_{n_2}(r_2s_2)...\psi_{n_N}(r_Ns_N)$$

$$\psi_a(r_1s_1, r_2s_2, ..., r_Ns_N) = \frac{1}{\sqrt{N!}} \sum_P (-1)^P P\psi_{n_1}(r_1s_1)\psi_{n_2}(r_2s_2)...\psi_{n_N}(r_Ns_N)$$

REMEMBER

The big news is that the antisymmetric wave function for N particles goes to zero if any two particles have the same quantum numbers ($n_i = n_j$, $i \neq j$). And that has a big effect in physics, as you see next.

It's not come one, come all: The Pauli exclusion principle

The antisymmetric wave function vanishes if any two particles in an N-particle system have the same quantum numbers. Because fermions are the type of particles that have antisymmetric wave functions, that's the equivalent of saying that in a system of N particles, no two fermions can have the same quantum numbers — that is, occupy the same state.

That idea, which Austrian physicist Wolfgang Pauli first formulated in 1925, is called the Pauli exclusion principle. The topic of discussion for physicists at that time was the atom, and the Pauli exclusion principle applied to the electrons (a type of fermion), which are present in all atoms.

The Pauli exclusion principle states that no two electrons can occupy the same quantum state inside a single atom. And that result is important for the structure of atoms. Instead of just piling on willy-nilly, electrons have to fill quantum states that aren't already taken. The same isn't true for bosons — for example, if you have a heap of alpha particles (bosons), they can all be in the same quantum state. Not so for fermions.

Electrons have various quantum numbers in an atom: n (the energy), ℓ (the angular momentum), m (the z component of the angular momentum), and m_s (the z component of spin). And using that information, you can construct the electron structure of atoms.

Figuring out the periodic table

One of the biggest successes of the Schrödinger equation, together with the Pauli exclusion principle (see the preceding section), is explaining the electron structure of atoms.

The electrons in an atom have a shell structure, with different shells defined by certain quantum numbers. Electrons fill that structure based on the Pauli exclusion principle, which maintains that no two electrons can have the same state:

>> **The major shells** are specified by the principal quantum number, n, corresponding to the distance of the electron from the nucleus.

>> **Shells, in turn, have subshells** based on the orbital angular momentum quantum number, ℓ.

>> **In turn, each subshell has subshells** — called orbitals — which are based on the z component of the angular momentum, m.

Noting the electron shell structure

So, each shell n has $n - 1$ subshells, corresponding to $\ell = 0, 1, 2, \ldots, n - 1$. And in turn, each subshell has $2\ell + 1$ orbitals, corresponding to $m = -1, -\ell + 1, \ldots, \ell - 1, \ell$.

Much as with the hydrogen atom, the various subshells ($\ell = 0, 1, 2, 3, 4$, and so on) are called the s, p, d, f, g, h, and so on states. So, for example, for a given n, an s state has one orbital ($m = 0$), a p state has three orbitals ($m = -1, 0$, and 1), a d state has five orbitals ($m = -2, -1, 0, 1$, and 2), and so on.

In addition, due to the z component of the spin, m_s, each orbital can contain two electrons — one with spin up, and one with spin down.

Filling up the electron shells

So how do electrons, as fermions, fill the structure of an atom? Electrons can't fill a quantum state that's already been taken. For atoms in the ground state, electrons fill the orbitals in order of increasing energy. As soon as all of a subshell's orbitals are filled, the next electron goes on to the next subshell; and when the subshell is filled, the next electron goes on to the next shell, and so on.

Of course, as you fill the different electron shells, subshells, and orbitals, you end up with a different electron structure. And because interactions between electrons form the basis of chemistry, as electrons fill the successive quantum levels in various atoms, you end up with different chemical properties for those atoms — which set up the period (row) and group (column) organization of the periodic table.

Giving Systems a Push: Perturbation Theory

Problems in quantum physics can become pretty tough pretty fast — which is another way of saying that, unfortunately, you just can't find exact solutions to many quantum physics problems. The lack of exact solutions particularly applies when you merge two kinds of systems. For example, you may know all about how square wells work (see Chapter 8) and all about how electrons in magnetic fields work (from classical physics; see Chapter 2 about Maxwell's equations), but what happens if you combine the two (square wells and magnetic fields)? The wave functions of each system, which you know exactly, are no longer applicable — you need some sort of mix instead. Perturbation theory comes to the rescue! This theory lets you handle mixes of situations, as long as the interference isn't too strong.

Introducing perturbation theory

REMEMBER

The idea behind *perturbation theory* is that you start with a known system — one whose wave functions you know and whose energy levels you know. Everything is all set up to this point. Then some new stimulus from the outside — a perturbation — comes along, disturbing the status quo. For example, you may apply an electrostatic or magnetic field to your known system, which changes that system somewhat.

Perturbation theory lets you handle situations like this — as long as the perturbation isn't too strong. In other words, if you apply a weak magnetic field to your known system, the energy levels will be mostly unchanged but with a correction. (*Note:* That's why it's called perturbation theory and not drastic-interference theory.) The change you make to the setup is slight enough so that you can calculate the resulting energy levels and wave functions as corrections to the fundamental energy levels and wave functions of the unperturbed system.

So what does it mean to talk of perturbations in physics terms? Say that you have this Hamiltonian:

$$H = H_0 + \lambda W \quad (\lambda \ll 1)$$

Here, H_0 is a known Hamiltonian, with known eigenfunctions and eigenvalues, and λW is the so-called perturbation Hamiltonian, where $\lambda \ll 1$ indicates that the perturbation Hamiltonian is small.

Finding the eigenstates of the Hamiltonian in this equation is what solving problems like this is all about — in other words, here's the problem you want to solve:

$$H|\psi_n\rangle = (H_0 + \lambda W)|\psi_n\rangle = E_n|\psi_n\rangle \quad (\lambda \ll 1)$$

The way you solve this equation depends on whether the exact, known solutions of H_0 are degenerate (that is, several states have the same energy) or nondegenerate. The next section solves the nondegenerate case.

Working with perturbations to nondegenerate Hamiltonians

Start with the case in which the unperturbed Hamiltonian, H_0, has nondegenerate solutions. That is, for every state $|\phi_n\rangle$, there's exactly one energy, En, that isn't the same as the energy for any other state: $H_0|\phi_n\rangle = E_n|\phi_n\rangle$ (just as a one-to-one function has only one x value for any y). You refer to these nondegenerate energy levels of the unperturbed Hamiltonian as $E^{(0)}_n$ to distinguish them from the corrections that the perturbation introduces, so the equation becomes

$$H_0|\phi_n\rangle = E^{(0)}_n|\phi_n\rangle$$

Note: From here on, I refer to the energy levels of the perturbed system as E_n.

The idea behind perturbation theory is that you can perform expansions based on the parameter λ (which is much, much less than 1) to find the wave functions and energy levels of the perturbed system. Here is the equation it leads to, up to terms in λ^2 in the expansions:

$$E_n = E_0 + \lambda\langle\phi_n|W|\phi_n\rangle + \lambda^2 \sum_{m \neq n} \frac{|\langle\phi_m|W|\phi_n\rangle|^2}{E^{(0)}_n - E^{(0)}_m} + \dots \quad (\lambda \ll 1)$$

That gives you the first- and second-order corrections to the energy, according to perturbation theory. You begin with the unperturbed energy E_0, modified by these two energy corrections.

For this equation to converge, the term in the summation must be small. And note in particular what happens to the expansion term if the energy levels are degenerate:

$$\frac{|\langle \phi_m \,|\, W \,|\, \phi_n \rangle|^2}{E^{(0)}_n - E^{(0)}_m}$$

TIP

The denominator here is the difference between the state energies. In a degenerate case, $n = m$ and you're going to end up with an $E^{(0)}_n$ that equals an $E^{(0)}_m$. This results in a denominator of zero, so the energy-corrections equation blows up (since you can't divide by zero). This approach to perturbation theory is no good if you have degenerate energies. You need a different approach to perturbation theory (coming up later in the section "Working with perturbations to degenerate Hamiltonians") to handle systems with degenerate energy states.

In the next section, I show you an example to make the idea of perturbing nondegenerate Hamiltonians more real.

Perturbation theory to the test: Harmonic oscillators in electric fields

Consider the case in which you have a small particle oscillating in a harmonic potential, back and forth, as Figure 15-5 shows.

FIGURE 15-5:
A harmonic
oscillator.

Here's the Hamiltonian for that particle, where the particle's mass is m, its location is x, and the angular frequency of the motion is ω:

$$H = \frac{-\hbar^2}{2m}\frac{d^2}{dx^2} + \frac{1}{2}m\omega^2 x^2$$

This Hamiltonian was covered in depth in Chapter 9, and you can refer to that content if you want to know how to use it to find exact solutions to quantum physics problems. And with perturbation theory, you can take that exact solution and use it as a springboard to get a new solution in a situation with slight changes from being perturbed.

Now assume that the particle is charged, with charge q, and that you apply a weak electric field, ε, as Figure 15-6 shows.

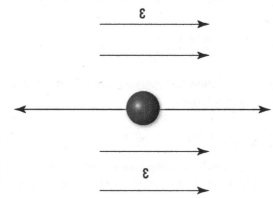

FIGURE 15-6:
Applying an
electric field to a
harmonic
oscillator.

The force due to the electric field in this case is the perturbation, and the Hamiltonian becomes

$$H = \frac{-\hbar^2}{2m}\frac{d^2}{dx^2} + \frac{1}{2}m\omega^2 x^2 + q\varepsilon x$$

The entire change between the Chapter 9 work and the work here comes from that final term ($q\varepsilon x$) representing the force of the weak electric field. Applying perturbation theory, the energy of the harmonic oscillator in the electric field turns out to be

$$E_n = \left(n + \frac{1}{2}\right)\hbar\omega - \frac{q^2\varepsilon^2}{2m\omega^2}$$

Fortunately, in this case it is also possible to calculate the exact result of the situation. When you do that, you find that the energy of the harmonic oscillator in both cases is the same! In other words, perturbation theory gives the same result as the exact answer.

REMEMBER

You can't always calculate the exact result, though, so sometimes you want to use perturbation theory to find the solution. Examples like this (for the harmonic oscillator) give evidence that perturbation theory, as an approach, is a way to match the exact solution (when you can calculate it).

Wave functions of the charged oscillator

So now you can look into the wave function for this charged oscillator (from the previous section's example). Here's the wave function of the perturbed system, to the first order:

$$|\psi_n\rangle = |\phi_n\rangle + \sum_{m \neq n} \frac{\langle \phi_m | \lambda W | \phi_n \rangle}{E^{(0)}_n - E^{(0)}_m} |\phi_m\rangle \ldots \quad (\lambda \ll 1)$$

Working through this equation and applying terms such as $\lambda W = q\varepsilon x$, you can translate the equation into this form:

$$|\psi_n\rangle = |n\rangle + \frac{q\varepsilon}{\hbar\omega}\sqrt{\frac{\hbar}{2m\omega}}\left(\sqrt{n}\,|n-1\rangle - \sqrt{n+1}\,|n+1\rangle\right)$$

Note: This equation means that adding an electric field to a quantum harmonic oscillator spreads the wave function of the harmonic oscillator.

Originally, the harmonic oscillator's wave function is just the standard harmonic oscillator wave function, $|\psi_n\rangle = |n\rangle$. Applying an electric field spreads the wave function, adding a component of $|n-1\rangle$, which is proportional to the electric field, ε, and the charge of the oscillator, q, like this:

$$|\psi_n\rangle = |n\rangle + \frac{q\varepsilon}{\hbar\omega}\sqrt{\frac{\hbar}{2m\omega}}\left(\sqrt{n}\,|n-1\rangle - \ldots\right)$$

And the wave function also spreads to the other adjacent state, $|n+1\rangle$, like this:

$$|\psi_n\rangle = |n\rangle + \frac{q\varepsilon}{\hbar\omega}\sqrt{\frac{\hbar}{2m\omega}}\left(\sqrt{n}\,|n-1\rangle - \sqrt{n+1}\,|n+1\rangle\right)$$

REMEMBER

You end up mixing states when you introduce a perturbation. That blending between states means that the perturbation you apply must be small with respect to the separation between unperturbed energy states, or you risk blurring the whole system to the point that you can't make any predictions about what's going to happen.

In any case, that's a nice result — blending the states in proportion to the strength of the electric field you apply — and it's typical of the result you get with perturbation theory.

TIP

Nondegenerate perturbation theory works and is strongly dependent on having the energy states separate so that your solution can blend them.

Working with perturbations to degenerate Hamiltonians

In the earlier section "Working with perturbations to nondegenerate Hamiltonians," I show that when you have two equivalent energy states, the second-order energy correction doesn't work. The denominator of that term becomes zero, which means the term itself becomes infinite and non-physical. This means that you must use a different approach to tackle systems with degenerate energies.

Since the system is *degenerate*, by definition, several states have the same energy. Suppose that you have a system with f states that have the same energy (so, you'd say the system is f-fold degenerate). That would result in this unperturbed Hamiltonian:

$$H_0 \left| \phi_{n_\alpha} \right\rangle = E^{(0)}_n \left| \phi_{n_\alpha} \right\rangle \quad (\alpha = 1,2,3,\ldots,f)$$

How does this affect the perturbation picture? The complete Hamiltonian, H, is made up of the original, unperturbed Hamiltonian, H_0, and the perturbation Hamiltonian, H_ρ:

$$H \left| \psi_n \right\rangle = \left(H_0 + H_\rho \right) \left| \psi_n \right\rangle = E_n \left| \psi_n \right\rangle$$

In zeroth-order approximation, you can write the eigenfunction $\left| \psi_n \right\rangle$ as a combination of the degenerate states $\left| \phi_{n_\alpha} \right\rangle$:

$$\left| \psi_n \right\rangle = \sum_{\alpha=1}^{f} a_\alpha \left| \phi_{n_\alpha} \right\rangle \ldots$$

The work involved in using this wave function is tedious, and beyond the scope of this book. (The end is near, after all!)

TIP

Here's one key point: Because these situations have degenerate energy states, more than one result will repeat identical energy solutions. In other words, when working with the vectors and matrices in degenerate problems like this, you find cases where multiple entries give you the same energy outputs. So the problems can look completely insurmountable one moment, and then several elements of the matrix collapse down to zero. So, you may be left with just a couple of calculations that you actually need to perform to reach the solutions. It doesn't always happen this way — but it's definitely nice when it does.

When Particles Collide: Scattering Theory

Many astounding discoveries of quantum physics throughout the twentieth century came from work at particle accelerators, where physicists can accelerate particles to immense speeds and then smash them together. It's a fun field of science, and when those particles get smashed together, analyzing what comes out of the collision offers a lot of new information. I discuss many related discoveries in Chapters 4 and 5, but in this chapter, I dive more deeply into how physicists make those discoveries.

Classically, you can predict the exact angle at which colliding objects will bounce off each other if the collision is elastic (that is, momentum and kinetic energy are both conserved). Quantum mechanically, however, you can only assign probabilities to the angles at which things scatter. And like other situations I cover in the book (for example, the double slit experiment from Chapter 3), instances that give exact classical answers yield only probabilities in quantum physics.

Introducing particle scattering and cross sections

In quantum physics, thinking in terms of the observables — what you can look at — is good. So, think of a scattering experiment in terms of particles in and particles out, like the illustration in Figure 15-7, for example. In the figure, particles are being sent in a stream (the incident particles) from the left and intersecting a target. Most of them continue on to the right unscattered, but some particles interact with the target and scatter.

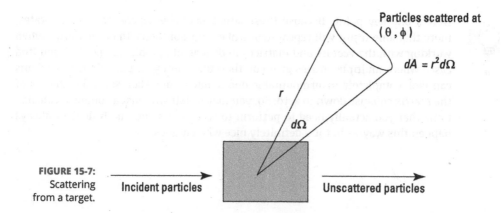

FIGURE 15-7: Scattering from a target.

Particles scattered at (θ, ϕ)

$dA = r^2 d\Omega$

r

$d\Omega$

Incident particles

Unscattered particles

The particles that scatter do so at a particular angle in three dimensions — that is, you give the scattering angle as a solid angle (depicted as a cone in Figure 15-7), $d\Omega$, which equals $\sin\theta\, d\theta\, d\phi$, where ϕ and θ are the spherical angles (see Chapter 13 for more on spherical angles).

The differential cross section is given by $\dfrac{d\sigma(\phi,\theta)}{d\Omega}$, and it's a measure of the number of particles per second scattered into $d\Omega$ per incoming flux. The incident flux, J (also called the current density), is the number of incident particles per unit area per unit time. So, $\dfrac{d\sigma(\phi,\theta)}{d\Omega}$ is

$$\frac{d\sigma(\phi,\theta)}{d\Omega} = \frac{1}{J}\frac{d\,N(\phi,\theta)}{d\Omega}$$

where $N(\phi, \theta)$ is the number of particles at angles ϕ and θ.

The differential cross section is the cross section for scattering to a specific solid angle. It has the dimensions of area (width and height), so calling it a cross section is appropriate. The total cross section, σ, is the cross section for scattering of any kind, through any angle. So, if the differential cross section for scattering to a particular solid angle is like the bull's eye, the total cross section corresponds to the whole target.

You can relate the total cross section to the differential cross section by integrating the following:

$$\sigma = \int \frac{d\sigma(\phi,\theta)d\Omega}{d\Omega} = \int_0^{2\pi}\int_0^{\pi} \frac{d\sigma(\phi,\theta)}{d\Omega}\sin\theta\, d\theta\, d\phi$$

Translating between the center-of-mass and lab frames

Getting into the details of scattering begins with a discussion of the center-of-mass frame versus the lab frame. Experiments take place in the lab frame, but you do scattering calculations in the center-of-mass frame, so you have to know how to translate your calculations between the two frames. I don't go all the way through the translations in this section, but I lay out the general way of thinking about them.

Figure 15-8 depicts scattering in the lab frame. One particle, traveling at $v_{1\,\text{lab}}$, is incident on another particle that's at rest ($v_{2\,\text{lab}} = 0$) and hits it. After the collision, the first particle is scattered at angle θ_1, traveling at $v_{1\,\text{lab}}$, and the other particle is scattered at angle θ_2 and velocity $v_{2\,\text{lab}}$.

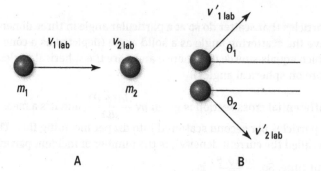

FIGURE 15-8:
Scattering in the
lab frame.

A   B

TIP

In the center-of-mass frame, the center of mass is stationary and the particles head toward each other. After they collide, they head away from each other at angles θ and π − θ. You use the center-of-mass frame for the calculations because the center of mass stays stationary, which means *the total momentum before and after the collision is zero in this frame*. This fact proves useful when performing the calculations. (It's always useful when total momentum is zero.)

You move back and forth between these two frames — the lab frame and the center-of-mass frame — to perform the translations. And you need to relate the velocities and angles (in a nonrelativistic way). Here are some starting relationships that are useful for relating the angles θ_1 and θ:

1. **Create equations for velocity.**

   The velocity of particle 1 in the lab frame equals the velocity of particle 1 ($v_{1\,lab}$) in the center-of-mass frame (v_{1c}) plus the velocity of the center of mass (v_{cm}) itself:

   Before collision: $v_{1\,lab} = v_{1\,c} + v_{cm}$

   After collision: $v'_{1\,lab} = v'_{1\,c} + v_{cm}$

2. **Apply the fact that the momentum is zero before and after the collision.**

   Before collision: $m_1 v_{1\,c} + m_2 v_{2\,c} = 0$

   After collision: $m_1 v'_{1\,c} + m_2 v'_{2\,c} = 0$

   Treat the collision as elastic, so kinetic energy is conserved.

   $$\frac{1}{2} m_1 v_{1\,c}^2 + \frac{1}{2} m_2 v_{2\,c}^2 = \frac{1}{2} m_1 v'^2_{1\,c} + \frac{1}{2} m_2 v'^2_{2\,c}$$

By building on these initial identities and relationships — and knowledge of trigonometry — you can find the different components of the velocity and write those components in terms of the angles and the masses, conveniently removing velocity from the consideration. If you know the masses of the particles and the angles of the output collisions, then you don't actually care about the velocities themselves.

Tracking the scattering amplitude of spinless particles

In this section, you get to look at the elastic scattering of two spinless nonrelativistic particles from the time-independent quantum physics point of view. Assume that the interaction between the particles depends only on their relative distance, $|r_1 - r_2|$. You can reduce problems of this kind to two decoupled problems (see Chapter 14 for details). The first decoupled equation treats the center of mass of the two particles as a free particle, and the second equation is for an effective particle of mass $\dfrac{m_1 m_2}{m_1 + m_2}$.

Starting with the effective particle-of-mass equation

The first decoupled equation, the free-particle equation of the center of mass, is of no interest to you in scattering discussions. The second equation is the one to concentrate on, where $\mu = \dfrac{m_1 m_2}{m_1 + m_2}$ gives you this Schrödinger equation to deal with:

$$\frac{-\hbar^2}{2\mu}\nabla^2\psi(r) + V(r)\psi(r) = E\psi(r)$$

You can use the preceding equation to solve for the probability that a particle is scattered into a solid angle $d\Omega$ — and you find this probability by the differential cross section, $\dfrac{d\sigma}{d\Omega}$.

REMEMBER

In quantum physics, wave packets represent particles. In terms of scattering, these wave packets must be wide enough so that the spreading that occurs during the scattering process is negligible (however, the wave packet can't be so spread that it encompasses the whole lab, including the particle detectors). Here's the crux: After the scattering, the wave function breaks up into two parts — an unscattered part and a scattered part. That's how scattering works in the quantum physics world.

Continuing with the scattered wave function

After the scattering of the spinless particles, the nonscattered wave function isn't of much interest to you, but the scattered wave function is. Here are the forms of the two wave functions.

>> **Incident wave function:** $\phi_{inc}(r) = Ae^{ik_0 \cdot r}$, where $k_0^2 = \dfrac{2\mu E_0}{\hbar^2}$

>> **Scattered wave function:** $\psi_{sc}(r) = Af(\phi, \theta)\dfrac{e^{ik \cdot r}}{r}$

For the scattered wave function, the $f(\phi, \theta)$ part is called the scattering amplitude. To put it simply, the scattering amplitude represents the portion of the wave that will be scattered. If you were analyzing a scattering problem, your job would be to find it. Here, A is a normalization factor and

$$k = \sqrt{\frac{2\mu E}{\hbar^2}}$$

where E is the energy of the scattered particle.

TIP

The scattering amplitude of spinless particles turns out to be crucial to understanding scattering from the quantum physics point of view. The problem of determining the differential cross section boils down to determining the scattering amplitude!

The Born approximation: Rescuing the wave equation

To find the scattering amplitude — and therefore the differential cross section — of spinless particles, you work on solving the Schrödinger equation:

$$\frac{-\hbar^2}{2\mu}\nabla^2 \psi(r) + V(r)\psi(r) = E\psi(r)$$

This eventually translates into this form (your scattering wave function), which, you may guess, is kind of a monster to try to solve exactly:

$$\psi(r) = \phi_{inc} - \frac{\mu}{2\pi\hbar^2}\int \frac{e^{ik|r - r'|}}{|r - r'|}V(r')\psi(r')d^3r'$$

But you can get close enough with a series of successive approximations, called the Born approximation (this is a famous result). I mentioned at the start of the chapter that physicists can earn their doctorates just by coming up with a method for how to solve a single quantum scenario. Max Born already had his doctorate when he came up with this method but, given that he recruited Heisenberg and some other key quantum physics pioneers, he well deserved to have *something* named after him! (You can find out more about Born in Chapters 3 and 16.)

To start, the zeroth-order Born approximation is just $\psi_0(r) = \phi_{inc}(r)$. And substituting this zeroth-order term, $\phi_{inc}(r)$, into the scattering wave function, in place of $\psi(r')$, gives you the first-order term:

$$\psi_1(r) = \phi_{inc}(r) - \frac{\mu}{2\pi\hbar^2}\int\frac{e^{i|r-r_1|}}{|r-r_1|}V(r_1)\phi_{inc}(r_1)d^3r_1$$

You then substitute this first-order approximation into the equation to get the second-order approximation:

$$\psi_2(r) =$$

$$\phi_{inc}(r) - \frac{\mu}{2\pi\hbar^2}\int\frac{e^{ik|r-r_2|}}{|r-r_2|}V(r_2)\phi_{inc}(r_2)d^3r_2$$

$$+ \frac{\mu^2}{4\pi^2\hbar^4}\int\frac{e^{ik|r-r_2|}}{|r-r_2|}V(r_2)\psi_1(r_2)d^3r_2\int\frac{e^{ik|r_2-r_1|}}{|r_2-r_1|}V(r_1)\phi_{inc}(r_1)d^3r_1$$

This nightmarish pattern continues for the higher terms, which you can find by plugging lower-order terms into higher ones. But after you pull together a Born approximation of the wave function, you have a starting point to define the wave function, and then you can use that to find the scattering amplitude and the differential cross section.

5

The Part of Tens

Discover more about ten of the most important quantum physics pioneers.

Dig deeper into ten of the biggest quantum physics triumphs.

Chapter 16

Ten Important Quantum Physics Pioneers

Throughout this book, I mention several of the most innovative and influential thinkers in the history of science, particularly those working in physics at the start of the twentieth century. These scientists transformed our understanding of the physical universe and the very fundamental nature of matter and energy — from the classical view to the quantum one described in this book.

In this chapter, I take some time to look more directly at these scientists. I provide a bit of biographical information and list some of their key accomplishments. Sometimes, I cover these accomplishments elsewhere in the book, but in some cases, I also introduce accomplishments that are outside the scope of this book's subject matter. And where appropriate, I offer information about interesting and popular biographies of these scientists for readers who might be interested in learning more.

Max Planck (1858–1947)

Max Karl Ernst Ludwig Planck was born in Kiel, Germany, into a family of intellectuals. During his college career, he studied in Berlin with physicists Hermann von Helmholtz and Gustav Kirchhoff, working on many of the key problems in thermodynamics.

Planck's most significant scientific accomplishment is the solution to the problem of black-body radiation, covered at length in Chapter 3, where he proposed the idea of quantifying the energy levels. This approach was intended as a mathematical simplification to allow for a solution, not to necessarily represent an actual physical restriction in nature. The physical constant that bears his name, Planck's constant, shows up everywhere in quantum physics. This work earned him the 1918 Nobel Prize in Physics.

Planck recognized the significance of Albert Einstein's special theory of relativity early on, and he became an avid supporter of both the theory and the young physicist who proposed it. Years later, both Planck and Einstein would join together in their skepticism of the implications of quantum physics, particularly as manifested in the Copenhagen interpretation (which explained quantum measurements as the collapse of the quantum wave function).

For a more comprehensive account of Max Planck, consider reading Brandon R. Brown's biography, *Planck: Driven by Vision, Broken by War* (Oxford University Press, 2015).

Albert Einstein (1879–1955)

Attempting to give a brief summation of Albert Einstein's life and work seems somehow disrespectful. I believe that no other intellectual figure in all of history has taken on such a profound, world-changing cultural perspective. In 1999, Albert Einstein was named *Time* magazine's Person of the Century.

While he is mostly known for his theory of relativity, Einstein was central to the creation and evolution of quantum physics at nearly every stage. Even with three decades of physics expertise, I found myself constantly surprised while writing this book by how often some thread of quantum physics ultimately traced back to Einstein himself. You can find a more comprehensive list of those influences in the online resources for this book (specifically, in the *Quantum Physics For Dummies*, Third Edition Cheat Sheet).

Niels Bohr (1885–1962)

Niels Bohr was born in Copenhagen, Denmark, and went on to become one of the central figures in quantum physics. His major accomplishment, which earned him the 1922 Nobel Prize in Physics, was recognizing that quantum theory could be applied to the structure of the atom. I cover this Chapter 3. The theory took the Rutherford model of the atom and applied Planck's quantum theory to explain (close enough) the results for hydrogen atom spectroscopy.

Springboarding off of this first success in quantum theory, Bohr became an enthusiastic cheerleader for the potential of incorporating quantum physics. Bohr went on to establish an institute for theoretical physics in Copenhagen, which became known as the Copenhagen Institute. Pretty much every European physicist of the era who worked on quantum theory eventually found their way to the Copenhagen Institute.

While Bohr was arguably not the lead innovator on any discovery (after his Bohr model of the atom), he had a profound impact on the generation of physicists that would transform nascent quantum theory into full mathematical quantum mechanics. He became a grandfatherly figure to the generation that followed.

Louis de Broglie (1892–1987)

Louis Victor Pierre Raymond, 7th Duc de Broglie, was born in Dieppe, France. From an aristocratic family, Louis began his studies with a love of history and assumed that he would pursue that field. But he eventually transitioned to physics.

Although he had a distinguished career, he is best known for his 1924 doctoral thesis, in which he predicted the wavelike behavior of matter (discussed in Chapter 3). This de Broglie hypothesis was experimentally confirmed in 1927, and ultimately earned him the 1929 Nobel Prize in Physics.

Werner Heisenberg (1901–1976)

Werner Karl Heisenberg was born in Würzburg, Bavaria. He came of age as a student of physics just as Niels Bohr was at the apex of his fame — founding the Copenhagen Institute, winning the Nobel Prize in Physics, and trying to convince everyone that if they could just figure out how quantum theory worked, then it would revolutionize physics. Heisenberg took him up on that challenge.

After meeting with Bohr in Copenhagen, Heisenberg developed the basic framework for what would become quantum mechanics. He sent a copy of his work to his brilliant friend Wolfgang Pauli and his mentor Max Born. Within a couple of weeks, Born submitted the results to a physics journal for publication, and enlisted another student, Pascual Jordan, to help work out the particulars.

Over a span of months in 1925, Heisenberg, Born, and Jordan worked out a matrix formulation of quantum mechanics, explaining the theory in precise, mathematical terms as tables of numbers that interact with each other. In 1927, Heisenberg published the uncertainty principle that bears his name; it is one of the most fundamental concepts of quantum physics. When the 1932 Nobel Prize in Physics was awarded for the development of quantum mechanics, it went to Heisenberg alone.

Although it is not a biography, Heisenberg's life is at the heart of the physics narrative, *Helgoland: Making Sense of the Quantum Revolution* (Riverhead Books, 2021) by Carlo Rovelli. Another good narrative approach to Heisenberg is the Michael Frayn play *Copenhagen* (1998), based around a real journey that Heisenberg made to Copenhagen in 1941 to meet with Niels Bohr. The two had a private conversation, but neither of them ever divulged the actual details of that conversation.

Erwin Schrödinger (1887–1961)

Erwin Rudolf Josef Alexander Schrödinger was born in Vienna, Austria-Hungary. At roughly the same time that Heisenberg, Born, and Jordan were developing their matrix formulation of quantum mechanics, Schrödinger developed the wave mechanics formulation. He also developed what has become known as the Schrödinger equation. The development of the Schrödinger equation earned him the 1933 Nobel Prize in Physics, which he shared with Paul Dirac.

Schrödinger's name is probably best known, both inside and outside of physics, for the thought experiment called Schrödinger's cat, which is intended to point out the absurdity of the Copenhagen interpretation of quantum mechanics. This thought experiment is covered in depth in Chapter 6.

Paul Dirac (1902–1984)

Paul Adrien Maurice Dirac was born in Bristol, England. While Heisenberg, Born, and Jordan were creating matrix mechanics, and while Schrödinger was creating wave mechanics, Paul Dirac was *also* devising the mathematical basis for quantum mechanics. His method matched that of the others who developed matrix mechanics, although he used an even more esoteric mathematical approach than theirs.

In 1930, Dirac published a book, *The Principles of Quantum Mechanics*, which introduced many of the mathematical tools used to explore quantum mechanics to this day, such as the Dirac notation (or bra-ket notation, as discussed in Chapter 7). He also worked to reconcile Einstein's general theory of relativity with the equations of quantum mechanics, and in doing so, he predicted the existence of antimatter, which was detected experimentally in 1932. For this work, Dirac shared the 1933 Nobel Prize in Physics with Erwin Schrödinger.

Max Born (1882–1970)

Max Born was born in Breslau, which is now known as Wroclaw, Poland. Born was the direct mentor to a number of revolutionary physicists who helped to create quantum mechanics. When Heisenberg was working on his formulation, it was Born who guided him toward thinking of the quantum observation values in a matrix format. And, ultimately, it was Born who recognized that Schrödinger's wave function could be understood to represent the probability of a given outcome. This was the major insight that led to him receiving the 1954 Nobel Prize in Physics.

Richard Feynman (1918–1988)

Richard Phillips Feynman was born in Queens, New York City, in the United States. Feynman earned a bachelor's degree in physics from the Massachusetts Institute of Technology, and then went on to Princeton University for his doctorate. In his 1942 dissertation, he laid out the groundwork for some fundamental ideas that he would return to in his research, including

>> The path integral formulation

>> Feynman diagrams

>> Positrons behaving like electrons moving backward in time

When the 1965 Nobel Prize in Physics was awarded for the development of quantum electrodynamics, it was shared by Feynman, Julian Schwinger, and Sin-Itiro Tomonaga, and by that point, these key contributions from Feynman were over 20 years old. Still, he was rightfully proud of them, famously having Feynman diagrams painted on his van.

Feynman wrote a number of books. Most of these are focused on physics, and are really just transcriptions of his lectures, although a couple are more autobiographical, namely *Surely You're Joking, Mr. Feynman: Adventures of a Curious Character* and *What Do You Care What Other People Think? Further Adventures of a Curious Character* (both from W.W. Norton & Company, 2018). For the definitive biography, look at *Genius: The Life and Science of Richard Feynman* (Pantheon, 1992), by James Gleick.

Murray Gell-Mann (1929–2019)

Murray Gell-Mann was born in Manhattan, New York City, in the United States. While Feynman was one of the last physicists to get a doctorate before the development of the nuclear bomb, Gell-Mann's 1948 bachelor's degree made him among the first generation of physicists to get their physics degree in the nuclear age. In the 1950s, he and Feynman worked together to help discover some fundamental aspects of the weak particle interaction.

Gell-Mann is best known as the founder of quantum chromodynamics (QCD), and specifically for being the source for the term *quark* to identify the new particles that made up hadrons (subatomic particles) such as the proton and neutron. (The term was a reference to a line in the James Joyce novel *Finnegans Wake*.) His research into the fundamental nature of particle physics earned Gell-Mann the 1969 Nobel Prize in Physics.

Chapter **17**

Ten Quantum Physics Triumphs

Quantum physics was created to explain physical measurements that classical physics couldn't explain. This chapter covers ten triumphs of quantum physics and highlights the key points you need to understand about each triumph that make it such a big deal. You can find more details about related concepts throughout the book and, particularly, in Part 1 and Part 2.

Wave-Particle Duality

Is that particle a wave? Or is that wave a particle? That's one of the questions that quantum physics was created to solve, because particles exhibited wavelike properties in the lab, whereas waves exhibited particle-like properties. The key moments in the history of developments associated with wave-particle duality are covered throughout the book, notably in Chapters 2, 3, and 4.

The Photoelectric Effect

One of the first successes of quantum physics involved explaining the photoelectric effect, as discussed in Chapter 3. The energy of electrons goes up with the frequency of the light, not its intensity; it's a situation that lends support to the light-as-a-stream-of-discrete-photons theory. Albert Einstein explained this in 1905 and eventually won the 1921 Nobel Prize in Physics for that explanation. The photoelectric effect has many practical applications, including in imaging and light detection. You can construct a material that emits different amounts of electrons as the light changes, creating a sensor that can detect changes in light or images. If you've ever walked up to a store and had the front door open for you, that technology has its roots in the photoelectric effect. Solar panels are another important technology that is descended from this discovery.

Postulating Spin

The Stern-Gerlach experiment results couldn't be explained without postulating spin, another triumph of quantum physics. This experiment sent electrons through a magnetic field, and the classical prediction was that the electron stream would create one spot of electrons on a screen. However, there were two spots, which corresponded to the two spins, up and down.

This detailed understanding of the behavior and structure of the electron, and of particles in general, led to a deeper examination of the fundamental nature of physical matter and its structure. The Standard Model of Particle Physics, the current theory that describes the basic building blocks of the universe, incorporates spin as a fundamental trait in defining all particles in the universe.

Differences between Newton's Laws and Quantum Physics

Perhaps the biggest triumph of quantum physics was building, in a span of about half a century, a comprehensive (and accurate!) theoretical and mathematical framework that provided an alternative to classical physics for analyzing matter and energy in the quantum realm. In classical physics, bound particles can have any energy or speed, but that's not true in quantum physics. And in classical physics, you can determine both the position and momentum of particles exactly, which isn't true in quantum physics (thanks to the Heisenberg uncertainty principle, which I discuss in Chapter 3). And in quantum physics, you can super-impose states on each other, and have particles tunnel into areas that would be classically impossible.

Heisenberg Uncertainty Principle

In his uncertainty principle, Heisenberg theorized that you can't simultaneously measure a particle's position and momentum exactly. This is one of the central theories that transformed classical physics and provided the basis for quantum physics. (More on this in Chapter 3.)

Quantum Tunneling

Probably one of the weirdest outcomes of quantum physics happens when a particle crosses a gap that it shouldn't (in classical terms) be able to cross. For example, how can an electron with energy E go into an electrostatic field (an example of the potential barriers referred to in Chapter 8), which you need to have more than energy E to penetrate? In classical physics, the answer is simple: It can't.

Quantum physics actually shows that in some cases, the particle may end up on the other side of this "insurmountable" gap. This process, called *quantum tunneling*, essentially has the particle jump across the gap without covering the intervening distance. The wave function is extremely small inside of the gap, inside of the electrostatic field. Since the square of the wave function represents the probability that a particle would be in this location, this means that you have an extremely small probability of finding the particle within the electrostatic field. There is an extremely low chance of finding the particle within the electrostatic field.

But the wave function on the other side of the gap from where the particle started is actually fairly significant by comparison. (See the work in Chapter 8 to confirm this.) In other words, contrary to all classical intuition, some significant probability exists that the particle will end up on the other side of the gap. In those cases, the particle has somehow gone from one side of this supposedly insurmountable barrier to the other, even though you'll almost never find these particles within the barrier! Since it doesn't have the energy to climb over the barrier, the metaphor is that the particle has tunneled through the barrier.

Discrete Spectra of Atoms

Science started the twentieth century still unsure about whether atoms actually existed. The debate, at that point, was largely among chemists. They knew that chemicals broke down into certain elements, but then didn't break down any further through chemical reactions. And the breakdown always happened in whole-number ratios, suggesting that the elements were maybe the smallest structure and couldn't be broken down any further.

Physicists jumped into the mix through thermodynamics and electromagnetic theory. They discovered that when you pump energy into atoms, you sometimes get the atoms to emit certain specific lines of light. The study of this electromagnetic spectrum of energy, sometimes called spectral lines, spawned its own branch of science, spectroscopy. This field of study goes all the way back to the 1860s, and so well predates quantum physics. Chemists and physicists at the start of the twentieth century adopted the idea that atoms were physical structures. The new field of quantum physics was fundamental for those studying spectroscopy and related to how matter and energy interacted.

Modeling the quantized nature of atoms and orbitals is another triumph of quantum physics. It turns out that electrons can't have any old energy in an atom but, instead, have only certain allowed quantized energy levels — and that was one of the foundations of quantum physics.

Harmonic Oscillator

The harmonic oscillator has all of Chapter 9 devoted to it, but it's worth mentioning here. Quantizing harmonic oscillators on the micro level was another triumph of quantum physics. Classically, harmonic oscillators can have any energy — but not quantum mechanically.

REMEMBER

Whenever a physicist is looking at a particle sticking around for any period of time, they're really looking at something that behaves as a quantum harmonic oscillator. Atoms in a lattice, or within a molecule, can be treated as oscillators. Particles trapped in energy wells or optical traps, including the basis of quantum computing, also rely on this understanding.

Square Wells

Like harmonic oscillators, quantizing particles bound in square wells at the micro level was another triumph for quantum physics. This is covered at length in Chapter 8, largely because it's one of those problems that quantum physics actually knows how to solve precisely. Classically, particles in square wells can have any energy, but quantum physics says you can have only certain allowed energies.

Schrödinger's Cat

Schrödinger's cat is a thought experiment that details some problems that arise in the macro world from thinking of the spin of electrons as completely non-determined until you measure them. For example, if you know the spin of one of a pair of newly created electrons, you know the other has to have the opposite spin. So, if you separate two electrons by light years and then measure the spin of one electron, does the other electron's spin suddenly snap to the opposite value — even at a distance that would take a signal from the first electron years to cover? Tricky stuff! The paradox at the heart of this thought experiment provides the basis for much of Chapter 6, so check it out for more details on this heated debate.

Index

A

A (amplitude)
  scattering, 299–300
  of waves, 19
abbreviating state vectors as kets, 107
adding
  time dependence to rectangular
    coordinates, 225
  time dependence to wave functions, 141–142
allowable energy states, 131
Ampere-Maxwell equation, 28
amplitude (A)
  scattering, 299–300
  of waves, 19
Andersonn, Philip W. (physicist), 280
angular momentum
  about, 181–182
  creating eigenstates, 186
  finding
    commutators, 184–185
    eigenvalues, 186–191
    eigenvalues of raising and lowering operators,
     191–192
  interpreting with matrices, 192–195
  Legendre differential equations, 195–196
  operators, 183–184
  setting up Hamiltonian equation, 182
  spherical coordinate system, 195–204
annihilation operator, 167
anode, 34
anticommutator, 117
anti-Hermitian operators, 117
antimatter, 76
antisymmetric wave functions, 281, 284–286
applying Hamiltonian operator to eigenstates, 166
Aspect, Alain (physicist), 99

assumptions, setting, 96–97
asymmetry, 280
atomic hypothesis, 31
atomic model, 50–52
atoms
  about, 31–34, 314
  components of, 65–67
  hydrogen and, 254–255
  multi-electron, 278–279

B

Balmer's equation, 51–52
baryon asymmetry, 280
baryonic matter, 280
baryons, 209
Bell, John (physicist), 98
Bell's inequality, 98
binding particles, in potential wells, 134–135
black body, 35, 42
black-body radiation
  about, 42–43
  spectrum, 44–45
Bohm, David (physicist), 93, 96
Bohr, Niels (physicist), 45, 50–52, 65, 72, 92, 254,
  264, 307, 308
Bohr radius, 139–140, 180, 264
Born, Max (physicist), 46, 58, 179, 300–301,
  308, 309
Born approximation, 300–301
bosons
  about, 73–74
  gauge, 79, 81
  Higgs, 80, 81, 210
  spin and, 208–210
bound state, 265
bra-ket notation, 106–111

interpretations
  about, 87–88, 91
  of angular momentum with matrices, 192–195
  Copenhagen, 92
  counterarguments for, 94–97
  debates about, 94–97
  entanglement, 90, 98–99
  hidden variables, 93–94
  many worlds, 92–93
  of measurement, 88–89
  of observer effect, 89–90
  of probability, 88–89
isotropic harmonic oscillators, 250–252

J

Jones, Andrew Zimmerman (author)
  String Theory For Dummies, 62
Jordan, Pascual (student), 179, 308

K

Kepler, Johannes (thinker), 14
kets
  abbreviating state vectors as, 107
  multiplying, 108–109
  relationships and, 110–111
  as state-less vectors, 109–110
kinetic energy, 16–18, 166
Kirchhoff, Gustav (physicist), 33, 42, 44, 306

L

lab frame, 297–299
Laplace, Pierre Simon (scientist), 28, 177–178
Laplacian operator, 112, 239
Large Hadron Collider (LHC), 81
laser technology, light and, 82
Law of Action and Reaction, 15
law of induction, 28
Law of Inertia, 15
law of universal gravitation, 14–15
Lawrence Livermore National Laboratory, 84

laws of motion, 14–15
laws of thermodynamics, 35
Legendre differential equations, 200–201
leptons, 209
LHC (Large Hadron Collider), 81
light
  electromagnetism and, 22–24
  laser technology and, 82
  as particles, 46–50
light amplification by stimulated emission of
    radiation (laser), 82
Linear Algebra For Dummies (Sterling), 103, 122
Linear momentum (P) operator, 112
linear operators, 114–115
longitudinal wave, 20
lowering operators
  finding eigenvalues of, 191–192
  matrices and, 194
  spin, 211–212
Lucretius, 31
luminiferous aether, 22
L_x, L_y, L_z, finding commutators of, 184–185

M

Manhattan Project, 62, 66, 83
many worlds interpretation, 92–93
many-particle systems, 276–286
The Mathematical Principles of Natural Philosophy
    (Newton), 14–15
matrices
  continuous representations compared with,
    126–129
  interpreting angular momentum with, 192–195
  mechanics of, 179–180
Maxwell, James Clerk (scientist), 27–28, 61
Maxwell's equations, 78–79
measurement
  interpretations of, 88–89
  uncertainties in, 38, 56
mechanics, classical, 14–18
Mendeleev, Dmitri (scientist), 32, 65
meson, 210

Michelson, Albert (physicist), 28–30, 60

microwave amplification by stimulated emission of radiation (maser), 82

Millikan, Robert Andrews (physicist), 45

modulated amplitude, 19

momentum
  about, 16–18
  angular
    about, 181–182
    creating eigenstates, 186
    finding commutators, 184–185
    finding eigenvalues, 186–191
    finding eigenvalues of raising and lowering operators, 191–192
    interpreting with matrices, 192–195
    Legendre differential equations, 195–196
    operators, 183–184
    setting up Hamiltonian equation, 182
    spherical coordinate system, 195–204
  position compared with, 55–57
  in quantum fields, 78

momentum operators, 183–184

momentum vectors, 109

Morley, Edward (physicist), 28–30, 60

motions, laws of, 14–15

multi-electron atoms, 278–279

multiplying bras and kets, 108–109

N

natural philosophers, 11

Neumann function, 244–245

neutrons
  about, 66–67
  in quantum chromodynamics, 80

Newton, Isaac (scientist), 14–15, 24, 28, 31, 312–313

Newton's laws, 312–313

Niels Bohr Institute, 52

Nobel Prizes in Physics, 45–46

nonlocal quantum potential, 94

nonlocality, 12

normalized vectors, 110

normalizing wave function, 140–141, 230–232

nuclear energy, 18, 83–84

nuclear fission, 83–84

nuclear forces, 67, 79–80

nuclear fusion, 84

nucleus, of atoms, 66–67

number operator, 168–169

numbers. *See* quantum numbers

O

observer effect
  about, 92
  interpretations of, 89–90

observing central potentials in 3D, 239–242

obstructions, to light, 22–23

operators
  about, 111–112
  for angular momentum, 183–184
  annihilation, 167
  anti-Hermitian, 117
  creation, 167
  exchange, 279–280, 285
  Gradient, 112
  Hamiltonian (H), 111, 164, 165–166, 276–278
  Hermitian, 115–116
  linear, 114–115
  lowering, 191–192, 194, 211–212
  momentum, 183–184
  number, 168–169
  position, 183–184
  raising, 191–192, 194, 211–212
  S^2, 213
  spin, 210–212
  S_z, 213–214
  unitary, 125–126
  unity or identity, 111

Opticks (Newton), 24

optics, 8

orbital angular momentum, 240

orbitals, hydrogen and, 272

orbits, of electrons, 51–52

Tip icon, 3
Tomonaga, Sin-Itiro (physicist), 46, 65, 310
total energy equation, finding, 224–225
total energy of a system, 165–166
transistors, 83
transmission coefficients
  about, 146
  finding, 150–151
transmissions, calculating probability of, 146–147
transverse wave, 20
trapping
  particles in infinite square potential wells, 136–143
  particles in potential wells, 134–136
trigonometry, 229
trough, of waves, 19
tunnelling, 152, 155–156, 313
two-particle systems, wave functions of, 286–287

U

Uhlenbeck, George E. (physicist), 207
uncertainties
  about, 8, 35–39
  applying to particle physics, 57
  in measurements, 56
  quantifying, 56–57
unitary operators, simplifying, 125–126
unity or identity (I) operator, 111
universal gravitation, law of, 14–15
unknowns, 35–39

V

variance, 38–39
vector field, 28, 78
vectors, state
  abbreviating as kets, 107
  about, 103–104
  basis of, 107
  bras and kets as basis-less, 109–110

comparing matrix and continuous representations, 126–129
creating in Hilbert space, 104–106
Dirac notation, 106–111
eigenvectors/eigenvalues, 120–124
finding
  anti-Hermitian operators, 117
  commutator, 116–117
  expectation values, 112–114
Heisenberg uncertainty principle, 117–120
Hermitian operators and adjoints, 115–116
kets and relationships, 110–111
linear operators, 114–115
multiplying bras and kets, 108–109
operators, 111–116
simplifying with unitary operators, 124–125
writing Hermitian conjugate as a bra, 108
von Helmholts, Hermann (physicist), 306

W

W (work function), 47
Warning icon, 3
wave function equation, finding, 136–139
wave functions
  about, 57–59, 127–129
  adding time dependence to, 141–142
  antisymmetric, 281, 284–286
  cubic potential, 233
  of harmonic oscillators, 294
  for hydrogen, 267–269
  normalizing, 140–141, 230–232
  particles and, 276–278, 285–288
  scattered, 300
  for spin states, 271
  symmetric, 281, 284–286
  of three-or-more particle systems, 288
  of two-particle systems, 286–287
wave mechanics, 179–180
wave number, 138
wave packet, 158–159

About the Author

Andrew Zimmerman Jones has a degree in physics from Wabash College. He is a member of the National Association of Science Writers, the National Science Teachers Association, and Mensa. He has a master's degree in mathematics education from Purdue University, as well as a certificate in applied game design from Central Michigan University. He has worked for over two decades in educational publishing by writing and editing math and science assessments and curricula at the state and national level, including with the Indiana Department of Education. For over ten years, he was the physics writer and editor at About.com, currently ThoughtCo. He is the lead co-author of *String Theory For Dummies*, wrote for NPR's NOVA Physics blog, and also scripted the TED-Ed video, "Does Time Exist?"

Dedication

To my beloved wife, Amber, for entangling her wave function with mine.

Author's Acknowledgments

Andrew would first like to give thanks to his wife, Amber, and sons, Elijah and Gideon, for being there to support him, even when he was driven frantic by deadlines, extensive revisions, and the vagaries of ordinary life. He would like to also offer thanks to all of the teachers, mentors, scientists, and authors who have taught him over the years. There are far too many to name them individually.

This book wouldn't be possible without the work of the late Dr. Steven Holzner, author of *Quantum Physics For Dummies*, Revised Edition. His extensive work on the earlier iterations of this book provided a solid foundation to build upon. Andrew would also like to offer his deepest appreciation to the wonderful editorial team that worked to pull this project together. He is thankful that Elizabeth Stilwell had the confidence in him to approach him with this opportunity and to trust in his recommendations on how to proceed with the project. It is impossible to express how much work the project editor, Leah Michael, and the copy editor, Marylouise Wiack, put into transforming wordy drafts of esoteric technical jargon into clear, concise explanations of these difficult concepts. They were truly a pleasure to work with at every moment.

Publisher's Acknowledgments

Acquisitions Editor: Elizabeth Stilwell, Associate Editor

Development Editor: Leah Michael

Copy Editor: Marylouise Wiack

Proofreader: Debbye Butler

Production Editor: Tamilmani Varadharaj

Cover Image: © agsandrew/Shutterstock

Take dummies with you everywhere you go!

Whether you are excited about e-books, want more from the web, must have your mobile apps, or are swept up in social media, dummies makes everything easier.

Find us online!

dummies.com

Leverage the power

Dummies is the global leader in the reference category and one of the most trusted and highly regarded brands in the world. No longer just focused on books, customers now have access to the dummies content they need in the format they want. Together we'll craft a solution that engages your customers, stands out from the competition, and helps you meet your goals.

Advertising & Sponsorships

Connect with an engaged audience on a powerful multimedia site, and position your message alongside expert how-to content. Dummies.com is a one-stop shop for free, online information and know-how curated by a team of experts.

- Targeted ads
- Video
- Email Marketing
- Microsites
- Sweepstakes sponsorship

20 MILLION PAGE VIEWS EVERY SINGLE MONTH

15 MILLION UNIQUE VISITORS PER MONTH

43% OF ALL VISITORS ACCESS THE SITE VIA THEIR MOBILE DEVICES

700,000 NEWSLETTER SUBSCRIPTIONS TO THE INBOXES OF

300,000 UNIQUE INDIVIDUALS EVERY WEEK

# of dummies

Custom Publishing

Reach a global audience in any language by creating a solution that will differentiate you from competitors, amplify your message, and encourage customers to make a buying decision.

- Apps
- Books
- eBooks
- Video
- Audio
- Webinars

Brand Licensing & Content

Leverage the strength of the world's most popular reference brand to reach new audiences and channels of distribution.

For more information, visit **dummies.com/biz**

PERSONAL ENRICHMENT

Staying Sharp
9781119187790
USA $26.00
CAN $31.99
UK £19.99

Facebook
9781119179030
USA $21.99
CAN $25.99
UK £16.99

Guitar
9781119293354
USA $24.99
CAN $29.99
UK £17.99

Investing
9781119293347
USA $22.99
CAN $27.99
UK £16.99

Beekeeping
9781119310068
USA $22.99
CAN $27.99
UK £16.99

Digital Photography
9781119235606
USA $24.99
CAN $29.99
UK £17.99

Meditation
9781119251163
USA $24.99
CAN $29.99
UK £17.99

Pregnancy
9781119235491
USA $26.99
CAN $31.99
UK £19.99

Samsung Galaxy S7
9781119279952
USA $24.99
CAN $29.99
UK £17.99

iPhone
9781119283133
USA $24.99
CAN $29.99
UK £17.99

Crocheting
9781119287117
USA $24.99
CAN $29.99
UK £16.99

Nutrition
9781119130246
USA $22.99
CAN $27.99
UK £16.99

PROFESSIONAL DEVELOPMENT

Windows 10
9781119311041
USA $24.99
CAN $29.99
UK £17.99

AutoCAD
9781119255796
USA $39.99
CAN $47.99
UK £27.99

Excel 2016
9781119293439
USA $26.99
CAN $31.99
UK £19.99

QuickBooks 2017
9781119281467
USA $26.99
CAN $31.99
UK £19.99

macOS Sierra
9781119280651
USA $29.99
CAN $35.99
UK £21.99

LinkedIn
9781119251132
USA $24.99
CAN $29.99
UK £17.99

Windows 10
9781119310563
USA $34.00
CAN $41.99
UK £24.99

SharePoint 2016
9781119181705
USA $29.99
CAN $35.99
UK £21.99

Fundamental Analysis
9781119263593
USA $26.99
CAN $31.99
UK £19.99

Networking
9781119257769
USA $29.99
CAN $35.99
UK £21.99

Office 2016
9781119293477
USA $26.99
CAN $31.99
UK £19.99

Office 365
9781119265313
USA $24.99
CAN $29.99
UK £17.99

Salesforce.com
9781119239314
USA $29.99
CAN $35.99
UK £21.99

Coding
9781119293323
USA $29.99
CAN $35.99
UK £21.99

dummies.com

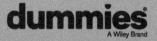
A Wiley Brand

Learning Made Easy

ACADEMIC

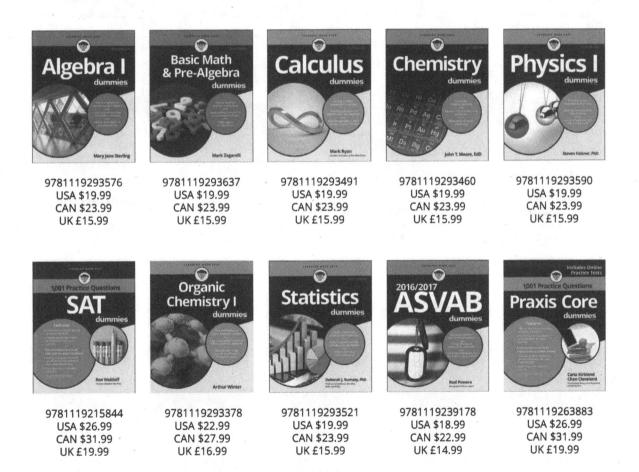

Algebra I dummies
Mary Jane Sterling
9781119293576
USA $19.99
CAN $23.99
UK £15.99

Basic Math & Pre-Algebra dummies
Mark Zegarelli
9781119293637
USA $19.99
CAN $23.99
UK £15.99

Calculus dummies
Mark Ryan
9781119293491
USA $19.99
CAN $23.99
UK £15.99

Chemistry dummies
John T. Moore, EdD
9781119293460
USA $19.99
CAN $23.99
UK £15.99

Physics I dummies
Steven Holzner, PhD
9781119293590
USA $19.99
CAN $23.99
UK £15.99

SAT dummies
1,001 Practice Questions
Ron Woldoff
9781119215844
USA $26.99
CAN $31.99
UK £19.99

Organic Chemistry I dummies
Arthur Winter
9781119293378
USA $22.99
CAN $27.99
UK £16.99

Statistics dummies
Deborah J. Rumsey, PhD
9781119293521
USA $19.99
CAN $23.99
UK £15.99

2016/2017 ASVAB dummies
Rod Powers
9781119239178
USA $18.99
CAN $22.99
UK £14.99

Praxis Core dummies
1,001 Practice Questions
Carla Kirkland
Chan Cleveland
9781119263883
USA $26.99
CAN $31.99
UK £19.99

Available Everywhere Books Are Sold

dummies.com

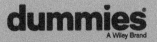

dummies
A Wiley Brand

Small books for big imaginations

GETTING STARTED WITH Coding
Get Creative with Code!
Camille McCue, PhD

9781119177173
USA $9.99
CAN $9.99
UK £8.99

MODDING Minecraft™
Build Your Own Minecraft Mods!
Sarah Guthals, PhD
Stephen Foster, PhD
Lindsey Handley, PhD

9781119177272
USA $9.99
CAN $9.99
UK £8.99

MAKING YouTube® VIDEOS
Star in Your Own Video!
Nick Willoughby

9781119177241
USA $9.99
CAN $9.99
UK £8.99

DESIGNING Digital Games
Create Games with Scratch™!
Derek Breen

9781119177210
USA $9.99
CAN $9.99
UK £8.99

GETTING STARTED WITH Raspberry Pi™
Program Your Raspberry Pi™!
Richard Wentk

9781119262657
USA $9.99
CAN $9.99
UK £6.99

EXPERIMENTING WITH Science
Think, Test, and Learn!

9781119291336
USA $9.99
CAN $9.99
UK £6.99

CREATING Digital Animations
Animate Stories with Scratch™!
Derek Breen

9781119233527
USA $9.99
CAN $9.99
UK £6.99

GETTING STARTED WITH Engineering
Think Like an Engineer!

9781119291220
USA $9.99
CAN $9.99
UK £6.99

WRITING Computer Code
Learn the Language of Computers!
Chris Minnick and Eva Holland

9781119177302
USA $9.99
CAN $9.99
UK £8.99

Unleash Their Creativity

dummies.com

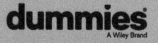